11 Progress in

Clinical Biochemistry and Medicine

Immobilized Enzymes in Medicine

By
V. P. Torchilin

With 59 Figures

Springer-Verlag
Berlin Heidelberg NewYork London Paris
Tokyo Hong Kong Barcelona

Ph.D. Dr. Sci. Vladimir P. Torchilin

Institute of Experimental Cardiology,
Cardiology Research Center of the USSR,
Academy of Medical Sciences,
3rd Cherepkovskaya Street 15A,
Moscow 121552, USSR

ISBN-13:978-3-642-75823-2 e-ISBN-13:978-3-642-75821-8
DOI: 10.1007/978-3-642-75821-8

Library of Congress Catalog Card Number 84-644052

Softcover reprint the hardcover 1st edition 1991

2151/3020-543210 – Printed on acid-free paper

Editorial Board

Table of Contents

1 Introduction

During the last two decades native biologically active compounds of protein origin, mainly enzymes, tend to turn from objects of fundamental studies into important instruments of technology in the broad sense of the word. Pure enzymes are already widely used in the food and pharmaceutical industries, for the synthesis of different chemical compounds and for analytical purposes. At last, enzymes turned out to be extremely effective drugs. In all cases, two principal factors are the main advantages of enzymes—their extremely high specificity, i.e. the ability to selectively influence one or few processes out of a great set of similar ones, and their high efficacy as catalysts; in other words, the ability to accelerate certain processes by thousand—or even million-times.

One of the most promising fields in the use of enzymes is in medicine. Today, when the mechanism of development of many human diseases is being elucidated at a molecular level, the role of endogenic enzymes in the organism is becoming clear; changes in their levels and activity determine the appearance and development of corresponding diseases. This is why enzyme therapy more and more attracts the attention of physicians, and enzymes themselves increasingly occupy an important place in the therapeutic arsenal of practical medicine. It is now possible to produce pure enzymes by relatively inexpensive methods in practically unlimited quantities and even to improve some of their properties.

The field of enzyme use in the medicine is not limited by their application as traditional drugs. Enzymes are also widely used in clinical-biochemical analysis to determine different metabolites or drugs in the blood, urea and other biological substance in body fluids and tissues. Furthermore, the development of enzyme immobilization methods, i.e. their coupling with different carriers, permits to obtain not only new drugs, but also numerous materials and systems for intracorporeal and extracorporal application—enzyme-containing films, bandages, dressing and drainage materials, columns for blood purification and different prosthetic devices.

In one book, it is impossible to present an exhaustive review and analysis of the innumerable reports dealing with the application of immobilized enzymes in medicine. This was not the aim of the author. His task was more modest—to acquaint the reader with the main problems in this field and the most important results. On some problems related to the theme of the book, there exists detailed and sufficient literature. First of all, this is true for the analytical application of immobilized enzymes in medicine, namely enzyme electrodes and enzyme immunoassays. Naturally, these chapters of the book will have a highly lapidary character, and the reader wishing to acquaint himself with these topics in detail should refer to special works. The author intends also to confine himself to basic data in sections devoted to the methods of immobilization and the resulting change of enzyme properties, because these questions are the main subject of numerous reviews and monographs on enzymes engineering. On the other hand, the author will pay great attention to therapeutic immobilized enzymes and the use of methods of immobilization to create systems of drug targeting in the organism.

2 Enzymes in Medicine: Advantages and Disadvantages

2.1 *Native Enzymes in Therapy and Clinical Analysis*

Before discussing particular problems connected with the use of enzymes in medicine, let us make some general statements. It is well known that enzymes are highly specific catalysts, facilitating appropriate reactions in subcellular compartments, cells, tissues and the organism as a whole.

It is quite natural that deviations from the normal functioning of one or several enzymes, which can result from some "errors" in their biosynthesis or the biosynthesis of their direct or indirect regulators, usually lead to disturbances in the normal homeostasis and the appearance of different diseases. Two questions arise: the first—how to influence the biosynthesis of a specific enzyme or its inhibitor in order to increase or to decrease its concentration in the affected tissue or in the whole organism; the second—is it possible to use exogenous enzymes or inhibitors in order to increase the concentration of the corresponding physiologically active compound in the organism. The current method now is the normalization of pathological shifts by the administration of exogenous enzymes or inhibitors.

Studies devoted to the use of enzymes as drugs are quite advanced and several books are already available in this field [1–3]. It is important to mention that during the last years, many new diseases have been described (usually inherited) connected with the deficiency of some lysosomal enzymes (so-called storage diseases) which can be treated only by the administration of exogenous enzymes [4–6]. Already, several approaches to enzyme therapy exist which can be divided (at least, conditionally) into several groups: (1) enzymes for replacement therapy, (usually digestive enzymes) which are used after surgery of the digestive tract, when the enzyme activity of the organism itself is not sufficient for effective food utilization; (2) the use of antitumor enzymes, usually possessing the specific ability to destroy some aminoacids, which are required for tumor growth; (3) enzymes for the treatment of inherited storage diseases (4) thrombolytic enzymes, which can act either directly—by lyzing thrombi, or indirectly—by activation of the fibrinolytic system of the blood; (5) enzymes active in some bacterial and viral infections; and (6) hydrolytic and antiinflammatory enzymes, for their action on pathologic or necrotic tissues.

The most evident and technologically simple is the use of digestive enzymes mixtures for replacement therapy. In this case, the drug is usually administered orally [1, 7], using capsules or tablets containing mammalian enzymes (mainly, proteases and lipases) mixed with stabilizers and fillers. In the stomach or in the intestine, enzymes release from tablets and can function long enough to provide complete digestion of a single food portion. Numerous preparations of the described type which are available now on the international pharmaceutical market, can be probably further improved by varying the ratio of the active components. This group also includes lactase, widely used in the treatment of lactose intolerance.

Antitumor enzymes are increasingly acquiring application in therapy. As it was already mentioned, their therapeutic action is based on their ability to decrease (via enzymatic break-down) the blood levels of nutrient essential for tumor growth. As a rule, these compounds are aminoacids. The principal requirements for these antitumor enzymes are formulated in [3]: (1) these enzymes should have low K_M value towards the appropriate substrate; (2) they should express maximal activity at physiological pH values; (3) and be sufficiently stable in the blood and other biological fluids; (4) their life-time in the circulation should be reasonably long; (5) they should not be inhibited by the reaction product even at high concentrations of the latter; (6) they should not require a cofactor or an easily dissociating prostetic group for their activity; (7) their preparations should contain minimal impurities such as endotoxins.

Among the antitumor enzymes, asparaginase is one of the most frequently used [8]. Asparaginase belongs to a group of wide-spread enzymes produced by both pro- and eucariotes. This enzyme hydrolyzes asparagine via desamination of the amino acid with the formation of aspartic acid:

$$\underset{\text{asparagin}}{H_2NOC{-}CH_2{-}\underset{\substack{|\\ NH_2}}{CH}{-}COOH} \xrightarrow[-NH_3]{\text{asparaginase}} \underset{\text{aspartic acid}}{HOOC{-}CH_2{-}\underset{\substack{|\\ NH_2}}{CH}{-}COOH}$$

The therapeutic action of asparaginase is based on the high requirement of some tumor cells for asparagine. For the first time, this effect was discovered in experiments with the administration of guinea-pig serum to animals with lymphoma. It turned out that only serum with sufficient asparaginase content suppressed tumor growth.

Later, asparaginase was used in the experimental treatment of different malignant diseases. The enzyme is active against acute lymphoblastic leukemia, but is ineffective against myeloid leukemia and other tumors. At the same time, the use of asparaginase in combination with other antitumor drugs sometimes potentiates the activity of the latter.

Asparaginase preparations used now in clinical practice are mainly of microbial origin. The enzyme sometimes is the only successful drug when no other antitumor agents can be used. At the same time, in some cases, the enzyme is not effective; remission, if it occurs, is only temporary; repeated asparaginase injections result in the development of intolerance; the clinical effect decreases or even disappears. Besides, the use of asparaginase is accompanied with different side-effects and complications: from moderate (weakness, headache, anorexia) to serious (anaphylactic shock, hepatic lesions).

The use of some other enzymes converting essential aminoacids also seems promising. Asparagine, alanine, arginine, cystein, glycine, ornithine, glutamine, glutamic acid, proline, serine, tyrosin and aspartic acid may be possible substrates for these enzymes [9]. Microbial glutaminase and glutaminase-asparaginase can be more or less successfully used to convert some of the above aminoacids. Unfortunately, in animal experiments and in clinical trials, these enzymes often demonstrate toxicity towards kidneys and the central nervous system. Cystein can be converted into pyruvate by cystein desulfatase; cystein aminotransferase and cysteine oxidase also can be used for therapeutic purposes. Excessive arginine can be destroyed by arginase, arginine deiminase and arginine decarboxylase. The review of experimental data concerning enzymatic degradation of essential aminoacids can be found in [9]. Some particular problems connected with enzyme therapy of cancer by degrading these aminoacids are also considered in [10]. An interesting approach to anticancer therapy suggested in [11] is the use of folate-degrading enzymes, which may inhibit tumor growth. On the other hand, some enzymes participating in folate degradation, first of all carboxypeptidase G_1, can decrease the toxicity of some important antitumor drugs like methotrexate. A pronounced anticancer effect was also demonstrated by different ribonucleases [12] and exonucleases [12a].

As already mentioned, an important area of enzyme therapy is the treatment of inherited diseases connected with the accumulation of particular toxic metabolites in cells due to the insufficient activity of endogenous intracellular enzymes [4–6]. Some of these diseases are listed in Table 1 together with the abnormal enzyme, accumulating product and clinical manifestation of each pathology. Gaucher's disease, e.g., is conditioned by the insufficiency of lysosomal glucocerebrosidase and leads to the increase in liver and spleen size, bone deformation and neurological disorders. The first manifestations of the disease appear usually in early age; patients are mentally and physically retarded. The only more or less successful attempt to treat the disease is connected with the use of purified exogenous enzyme applied intravenously [6]. One more serious lysosomal storage disease is Pompe's disease, the result of α-1,4-glucosidase deficiency. The accumulation of glycogen in lysosomes causes muscle dystrophy and child death in 1–2 months after birth. The success in the treatment of another inherited disease—phenylketonuria—is rather limited and connected mainly with dietotherapy limiting the quantity of phenylalanine in the food. The only reasonable alternative seems to be the treatment with phenylalanine ammonia lyase.

Enzyme therapy can be also useful for the treatment of the most often inherited hemostatic disturbance—hemophilia A [13]. This disease is caused by the deficiency of blood clotting factor VIII, which results in frequent and

Table 1. The main types of the inherited enzyme deficiencies

Disease	Enzyme deficiency	Main accumulated substance	Principal clinical manifestations
Glycogenosis Type II (Pompe's disease)	α-1,4-Glucosidase (acid maltase)	Glycogen	Increasing myopathic infections, heart failure
G_{M1} Gangliosidosis	G_{M1}-β-Galactosidase	G_{M1}-Ganglioside	Neurologic disorders
G_{M2} Gangliosidosis (Tay-Sachs' disease)	β-N-acetylgexoaminidase A	G_{M2}-Ganglioside	Neurologic disorders, adynamia, blindness, emanciation
Ceryngolicydosis (Gaucher's disease)	β-Glucosidase (glucocerebrosidase)	Glucocerebroside	Bone deformation, splenohepatomegalia, neurologic disorders, emnciation
Mannosidosis	α-Mannosidase	Mannose (glycogen)	Mental and neurologic disorders, loss of vision
Red blood cells enzymopathy	Glucose-6-phosphate-dehydrohehepase		Acute and chronic anemia, hemolytic icterus
Phenylketonuria	Phenylalanine-hydroxylase	Phenylalanine	Imbecility, neurologic disorder
Hemophilia	Factor VIII (antihemophilic factor)		Bleeding, secondary anemia

prolonged bleeding. Usually, the transfusion of "normal" blood (containing the whole set of blood clotting cascade enzymes) is used for the treatment of hemophilia. The use of pure factor VIII seems to be even more rational.

Porphyriases caused by the accumulation of different intermediates of heme biosynthesis in liver cells form another big group of inborn enzyme pathologies. One of them is connected to the excessive activity of aminolevulinic acid synthetase. This activity can be determined by both the inherited increase in enzyme biosynthesis and the inherited decrease in the activity of uroporphyrine synthetase. The latter causes the increase in the aminolevulinic acid synthetase activity via a feedback mechanism. One can eliminate liver accumulation of porphyrines either by the inhibition of enzymes catalyzing the initial stages of porphyrine biosynthesis or by the administration of enzymes catalyzing further transformations of aminolevulinic acid, e.g., gamma-aminolevulinic acid dehydratase.

Detailed data concerning different genetic storage disease including the discussion of their mechanisms and the strategy of enzyme therapy are presented in [6]. Table 2 (taken from [6] with some changes) describes the main storage diseases, abnormal enzymes and affected cells. It is also pointed out in [6] that for many storage pathologies, the treatment with purified exogenous enzymes gives

Table 2. Primary cellular site of pathology in the lysosomal diseases [adapted from (6)]

Disease	Defective enzyme	Primary site of lysosomal pathology
Lipidoses		
Fabry's disease	α-Galactosidase A	Vascular endothelium
Gaucher's disease	β-Glucosidase	RES[a], Neuron
G_{M2} Gangliosidosis	β-Hexosaminidase A and B	Neuron, RES
G_{M1} Gangliosidosis	G_{M1}-β-Galactosidase	Neuron, RES
Krabbe's disease	Galactosylceramide: β-galactosidase	Neuron
Metachromatic leucodystrophy	Cerebroside sulfatase	Schwann and glial cells
Nieman–Pick's disease	Sphingomyelinase	Neuron, RES
Farber's disease	Ceramidase	Neuron, RES
Mucopolysaccharidoses		
MPS I	α-L-Iduronidase	Neuron, fibroblasts
MPS II	α-L-Iduronidate sulfatase	Neuron, fibroblasts
MPS III		Neuron
A	Heparan N-sulfatase	
B	N-acetyl-α-D-glucosaminidase	
C	Acetyl CoA: α-glucosaminide N-acetyl transferase	
MPS IV	Hexosamine 6-sulfatase	Chondrocytes, fibroblasts
MPS VI	N-Acetylgalactosamine-4-sulfatase	RES, fibroblasts
MPS VII	β-Glucuronidase	Neuron, fibroblasts
Glycoproteinoses		
Mannosidosis	α-Mannosidase A and B	Neuron
Fucosidosis	α-L-Fucosidase	Neuron
Aspartylglucosaminuria	Aspartylglucosaminidase	Neuron
Sialidosis	Neuraminidase	RES, fibroblasts
Glycogenosis		
T100 II	Acid α-glucosidase	Striated muscle cell

a-RES = Reticuloendothelial system.

very promising results. Unfortunately, therapy with native enzymes gives only short-term relief because their half-life in circulation is usually not more than 20–30 minutes [3].

One of the most developed areas of enzyme therapy is the use of enzymes for the treatment of different abnormalities in the blood clotting system (14–18). At present this type of therapy is based on the knowledge of molecular mechanisms

Table 3. Enzymes with potential for medical therapy (a partial list). (From Klein M.D., and Langer R., TIBTECH-July, 1986, 179, with permission)

Enzyme (source)	Disease	Enzyme action
α-1,4-Glucosidase (Aspergillus niger)	Type II glycogenosis (Pompe's disease)	Metabolizes glycogen to allow its removal from cells
β-Galactosidase (figs)	Fabry's disease (heart and kidney failure due to ceramide accumulation)	Removes terminal galactose from galactosylgalactosyl-glucosylceramide in cell lysosomes
Arginase (rat liver)	Familial hyperargininemia	Removes arginine from blood
Asparaginase (E. coli)	Cancer: especially acute lymphocytic leukemia	Destroys asparagine, an essential amino acid for tumor growth
β-Glucuronidase (bovine liver)	Mucopolysaccharidosis VII	Metabolizes mucopolysaccharides so they do not accumulate
Bilirubin oxidase (Myrothecium verrucaria)	Jaundice; liver disease	Converts bilirubin to biliverdin
Carboxypeptidase G 1 (Pseudomonas stutzeri)	Tumors treated with methotrexate, e.g. osteogenic	Antagonizes mathotrexate; inhibits folate metabolism
Catalase (cows)	Acatalasemia; ? treatment of wide spectrum of diseases thought to be mediated by oxygen free radicals	Reduces H_2O_2 to water; inhibits generation of hydroxyl radicals
Fibrinolysin	Thromboembolic occlusive vascular disease	Clot dissolution
Glutaminase (Pseudomonas putrefaciens)	Cancer	Destroys glutamine, an essential amino acid for tumor growth
Heparinase (Flavobacterium heparinum)	Extracorporeal circulation	Inactivates heparin by removing disaccharides
Liver microsomal enzymes (rabbit liver)	Liver failure	Various detoxification steps including glucuronidation of phenols
Phenylalanine ammonia lyase (Rhodotorula glutinis)	Phenylketonuria	Metabolizes phenylalanine to transcinnamic acid and ammonia
Streptokinase (Streptococcus sp.)	Thromboembolic occlusive vascular disease	Activates plasminogen in blood which lyses clots
Superoxide dismutase (bovine liver)	Many inflammatory diseases thought to be mediated by oxygen free radicals	Converts oxygen free radicals to H_2O_2 and O_2
Terrilythin	Peritonitis	Improved survival in fecal peritonitis in rabbits
Tyrosinase (mushrooms)	Liver failure	Metabolizes tyrosine and free phenols
UDP Glucuronyl transferase (rabbit liver)	Jaundice; liver disease	Conjugates bilirubin and other toxins to water soluble form
Urea cycle enzymes	Liver failure	Metabolizes ammonia
Urease	Renal failure	Converts urea to ammonia
Uricase (hog liver)	Hyperuricemia due to gout	Rabbit anti-uricase serum inhibited free but not immobilized uricase
Urokinase (human urine)	Thromboembolic occlusive vascular disease	Activates plasminogen in blood which lyses clots

of fibrinolysis regulation [19]. Thrombolytic enzyme therapy uses many different enzymes—fibrinolysin (plasmin), streptokinase, urokinase (pro-urokinase), tissue plasminogen activator [20]. The market of thrombolytic enzymes is growing continuously and new methods of their application are introduced into clinical practice. New modifications of thrombolytic enzymes are being developed such as single-chain urokinase [21]. Different therapeutic enzyme mixtures have been suggested; thus, the simultaneous action of hialuronidase and urokinase on experimental miocardial infarction has been studied [22]. Thrombolytics obtained by genetic engineering are already in clinical trials [23].

It was found experimentally that reperfusion can lead to toxic tissue injury from free radical oxygen derivatives. This observation has led to the idea of using free radical detoxicating enzymes (like superoxide dismutase and catalase) as a part of thrombolytic therapy [24].

Many other enzymes are also being considered as promising therapeutic agents. Thus, elastase can be used in the treatment of arthritis [25]; pronase is used for the treatment of spleen and liver diseases [26]; collagenase and its mixture with hialuronidase are promising in the treatment of spleen and liver diseases [26]; collagenase and its mixture with hialuronidase are considered for the treatment of postoperative scars [27]; mixtures of proteases with ribonuclease and deoxyribonuclease showed good results in the prevention and treatment of postoperative pneumonia [27]; lysozyme is highly effective against viral diseases, including hepatitis, and in stomatology for caries treatment [27]. There are also several more exotic but not less interesting examples [28–30].

The use of enzymes for the treatment of viral diseases, different inflammations, wounds and burns is reviewed in [1–3]. In these situations, enzymes are used as components of different ointments and powders. The most important examples of therapeutic applications of enzymes are summarized in Table 3.

2.2 Difficulties Connected with Enzyme Application in Medicine

A natural question arises: why are enzymes which possess so many advantages from the biomedical point of view, still used so rarely in everyday clinical practice?

The reasons of this situation are serious enough and they can be conditionally divided into two principal groups. First, there are difficulties of economic character—the production of pure enzymes is still limited and they are relatively expensive, especially enzymes obtained from animal sources. Second, there are factors connected with the nature of enzymes as complex protein macromolecules, which are, as a rule, foreign to the recipient organism. This leads to low stability of the majority of enzymes at physiological pH and temperatures, particularly when these enzymes have to be active in other than their normal environment. As a result, exogenic enzymes being administered into the organism undergo numerous changes due to external influences (schematically shown in Fig. 1), which lead to their rapid inactivation. All this leads to the necessity of

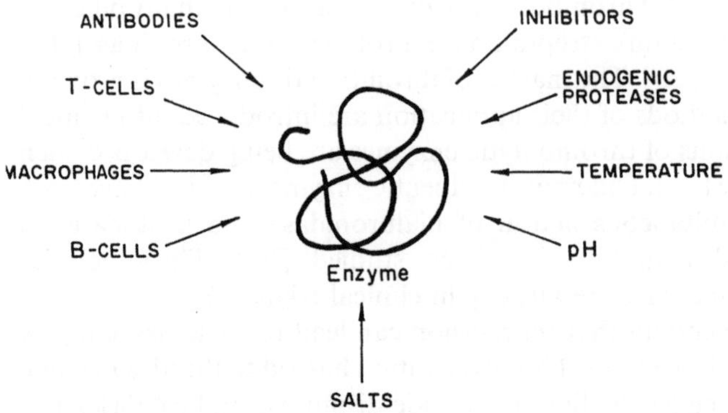

Fig. 1. Factors affecting an enzyme in the organism

frequent repeated administrations of enzymes. This necessity is also conditioned by the fact that pure enzymes normally do not possess a pronounced affinity for the lesion. As a result, in order to create a high local concentration of a therapeutic enzyme, it is necessary to obtain a high concentration in the systemic blood flow or in the organism as a whole. It is also evident that a foreign protein, bristling with different antigenic determinants, will interact with immunocompetent cells in the organism and cause immune reactions, which very often make its further application impossible. That is why researchers have to find ways which permit, on the basis of the available enzymes, to obtain certain derivatives with unaltered therapeutic activity, but deprived of toxicity and antigenicity.

To improve the prospects of enzyme immobilization, it is necessary to alter their insufficient storage stability. Even at low temperatures and in the dark, many enzymatic drugs can not be stored too long.

Enzyme modification (immobilization) is expected to solve one of the difficulties, which makes the parenteral application of enzymes limited and in some cases even impossible. Besides allergic complications mentioned above, many enzymes cause pronounced local irritation. A harmful action can be of systemic character, manifested as the injury to blood or other organs (for example, hemolysis upon the administration of proteolytic enzymes into the blood flow). For some enzymes, the difference between a therapeutic and toxic dose is very small. Enzyme modification (immobilization) capable to selectively reduce the enzyme contact with cell receptors seems to be the most realistic approach to decrease the harmful effect of such perparations.

2.3 How Immobilization Can Help

Immobilization is the physical (adsorptional) or chemical (covalent) binding of an enzyme with a carrier matrix (soluble or insoluble). This matrix protects immobilized protein molecules against different inactivating influences. At the same time, it causes a minimal effect on the normal state of enzyme active sites. How can this procedure solve the problem or facilitate the use of therapeutical use and increase enzyme efficacy in clinical situations?

Evidently, immobilized enzymes are expected to possess a set of properties which are absent in their native precursors. In the simplest case, it can be just an increase in enzyme storage stability, which is a complex and important task. The mechanisms of enzyme inactivation in the organism are a part of a natural reponse of the organism of any foreign protein (see Fig. 1). They include powerful inhibitory systems of the blood and tissues which becomes the major obstacle for the parenteral enzyme application. Thus, the clear aim of therapeutic enzyme immobilization or modification is the diminution of enzyme affinity to natural inhibitors.

A very important point is also the immune response of the macroorganism to foreign proteins. Specific antibodies against a given protein can rise during life or as a result of repeated administration of a therapeutic protein (enzyme). In any case it leads to enzyme inactivation or even to allergic reactions (as side effect). Thus, enzyme immobilization should decrease its immunogenicity and affinity to specific antibodies.

Parenterally administered enzymes should exist in the active form for a long period of time. It should also reach and penetrate an affected tissue. Improved pharmacokinetic properties of an enzyme and an increase in bioavailability can be achieved via enzyme modification, creation of "targeted" enzyme preparations or the use of immobilized enzymes in some extracorporeal systems which can temporarily replace affected organs.

Besides the improvement of biological properties of an enzyme, it is also desirable to get an enzyme derivative which is easier to handle. It is known that many enzymes (thrombolytics among them) can be administered to a patient only in the form of a prolonged drip infusion. It automatically transfers the start of the

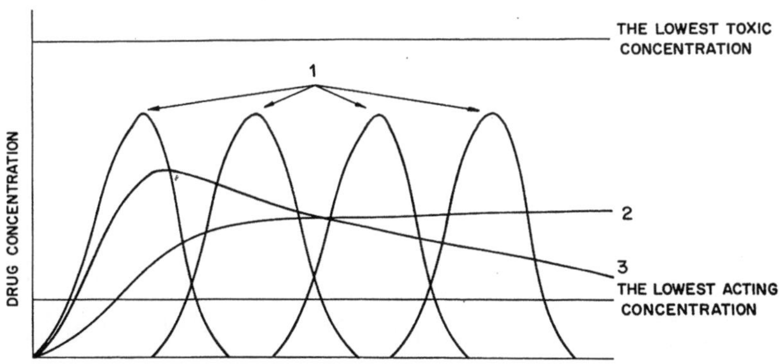

Fig. 2. Drug concentration in plasma: (*1*) routine periodic administration; (*2*) steady-state drug delivery after administration of immobilized drug; (*3*) prolonged drug delivery with gradually decreased release rate

therapy to the manifestation of the clinical conditions. An immobilized enzyme which possess an increased stability and prolonged half-life could be administered as a single injection of the whole therapeutic dose. After that, the activity of the appropriate enzyme in the body is supposed to be maintained on the required level until the next injection, after several hours or even days (see Fig. 2, curves 2 and 3).

Let us ennumerate what can we expect from immobilized therapeutic enzymes: (1) increased stability and prolonged activity in the body; (2) decreased antigenicity, which permits us to administer repeated therapy; (3) decreased affinity to natural inhibitors; (4) the possibility to administer the whole therapeutic dose of an enzyme in a single injection, therapeutic procedure which is more simple and convenient both for the patient and the physician; (5) a decrease in the total quantity of an enzyme needed for the treatment, which makes the treatment more economical.

Also in biochemical analysis the use of immobilized enzymes should make the whole analytical procedure more economical, simple, and faster. Thus, when using the native enzyme, each single analysis requires the irreversible expenditure of a certain enzyme quantity, which can not be separated from the reaction mixture. The use of reactions with immobilized (insolubilized) enzymes allows us to perform many analyses with a single dose of an enzyme. The possibility of combining several immobilized enzyme reactions into one system will permit the determination of several parameters using a single biological sample. As we shall see later, the use of immobilized enzymes leads to the creation of completely new and very efficient analytical methods.

3 Immobilization of Therapeutic Enzymes

Independent of the application, all enzymes share many similar properties. Evidently, mechanisms of their inactivation and stabilization also should have a lot of similarities. Peculiarities of the further use of immobilized enzyme are important mainly for the proper choice of optimal carriers and immobilization methods in each case. So, in the first section of this chapter we will discuss the most general principles of enzyme stabilization and immobilization.

3.1 General Principles of Enzyme Stabilization and Immobilization

During the last 20 years the problem of enzyme stabilization and immobilization have been discussed in many original publications, reviews and monographs [32–38]. Without going into details of the phenomena occurring during enzyme inactivation and stabilization and the kinetic description of corresponding processes, we will consider only the most important data [39, 40].

Quantitative characterization of enzyme stability. Enzyme stability, i.e. the ability to perserve its catalytic activity at the given conditions (pH, temperature, solution composition), is determined by both enzyme structure and the nature of its interaction with the environment in a solution, on a carrier surface or in a polymeric gel. Different parameters are used to characterize enzyme stability quantitatively. For this purpose many authors use the rate constant of monomolecular enzyme inactivation K_{in}. Unfortunately, very often enzyme inactivation does not follow first-order kinetics—an enzyme preparation can be heterogeneous and contain several fractions, with different K_{in} values, or enzyme inactivation mechanisms can be complex and consist of several consecutive or parallel steps. To overcome this, different empirical parameters are used to describe enzyme stability:

1. The rate constant (or the rate) of enzyme inactivation at the given conversion degree (for example, 50 or 75 per cent);
2. The time interval, during which enzyme activity decreases by 2-fold (i.e. enzyme half-life);

3. The fraction of enzyme residual activity after its incubation during a fixed time interval in the presence of a denaturating agent or at a given temperature.

To characterize enzyme preparations used for practical purposes, one need to know their storage stability and stability in operating conditions.

The different processes leading to the inactivation of enzymes and deterioration of their active centers can be divided into two large groups:

1) physical (or physico-chemical) and
2) chemical inactivation processes.

The first group includes: a) conformational intramolecular protein transformation from the effect of temperature, pH, high salt concentration, organic solvents or detergents; b) the dissociation of enzyme subunits or enzyme-cofactor complexes and the association of enzyme molecules with the formation of inactive associates; c) enzyme non-covalent complexing with ions or low molecular weight and high molecular weight compounds, affecting the native structure of the active site.

The chemical inactivation of enzymes includes both intermolecular interactions of enzymes and the modification of enzymes by different compounds in solution. The latter includes the oxidation of SH-groups in sulfhydryl enzymes and Fe (II) atoms in heme-containing proteins by atmospheric oxygen; redistribution of SH- and S—S-bonds (thiol-disulfide exchange); destruction of labile side-groups like tryptophan and metionine. In all cases mentioned above, the modification of two types of enzyme reactive groups is really crucial: groups which compose the active site of an enzyme and groups which are important for the spatial organization of the enzyme molecule and the stabilization of the active site structure.

One of the most general principles of enzyme stabilization is the limitation of polypeptide chain conformational mobility. As a result, a drastic increase in enzyme stability can be achieved by enzyme entrapment into concentrated polymeric gels, via formation of multiple covalent or non-covalent bonds between an enzyme molecule and a carrier matrix, or following the incorporation of intramolecular cross-links into protein molecules.

Another stabilization method is the prevention of dissociation or association processes in the enzyme molecule. The covalent immobilization of enzymes on a carrier matrix is a good way to eliminate association processes. At the same time, dissociation processes can be affected by the carrier matrix in different ways. In many cases, the dissociation rate of active associates does not change upon the immobilization of an oligomeric enzyme. Nevertheless, the relative content of active associates should be generally higher for an enzyme in the immobilized state than in solution, because dissociated subunits of the immobilized enzyme remain close to each other, which leads to an increase in the effective association rate.

The protection of an enzyme against denaturating agents (pH, organic solvents, metal ions, oxygen) can be achieved by different ways. Two most general approaches are: a) the steric separation of an enzyme and denaturating agent, and

b) the modification of an enzyme which hinders its interaction with denaturating factors.

Sometimes, enzyme stabilization can be observed in the presence of different organic and inorganic compounds [36, 39]. Thus, bivalent cations (Ca^{2+}, Mg^{2+}, Zn^{2+}) often stabilize enzymes. Possible mechanisms of stabilization are the appearance of ionic bridges between protein carboxylic groups and the increase in intraglobular hydrophobic interactions due to the decrease of the summary charge of the protein globule resulting from ionic interactions. The increase in hydrophobic interactions inside the protein globule resulting from concentrated salts leads to the formation of more compact enzyme conformations and is the main reason of enzyme stabilization under the conditions discussed [41].

An enzyme can be also stabilized by complexing with coenzymes, substrates and inhibitors.

Organic compounds—polyoles, carbohydrates, some solvents—can stabilize enzymes due to the formation of protein:organic compound complexes, the development of hydrogen bonding and changes in water structure in the microenvironment of a protein.

In many cases, natural and synthetic polymers (albumin, polyvinylpyrrolidone and other vinyl polymers, polysaccharides, polyethyleneimine) proved to be effective enzyme stabilizers. The stabilization of enzymes by polyelectrolytes or other proteins can be explained in different ways. The formation of enzyme:polyelectrolyte complexes changes the micro-pH value in the vicinity of the protein globule and the macroscopic pH-dependence of enzyme stability. On the other hand, the local concentration of denaturating agents can be drastically decreased inside such complexes. It is known, for example, that polycations can protect enzymes from the inactivation by heavy metal ions, preventing the latter from the contact with protein SH-groups.

The formation of specific enzyme:polymer complexes stabilizes the active enzyme structure because new bonds (hydrogen, electrostatic, donor-acceptor) develop between the enzyme and a carrier matrix.

Similar stabilization mechanisms can be discovered when enzymes are included into insoluble polymeric gels (neutral or charged) or non-covalently immobilized on different solid carriers, i.e., adsorbed on silica gel, cellulose, ion-exchange resins and other organic and inorganic carriers.

Stabilization effects are usually very effective when polymeric carriers, capable of formation of additional non-covalent interactions, are used with enzyme molecules.

The stabilizing action of cross-linked ion-exchangers on enzymes includes at least two different factors: a) a kinetic one (the redistribution of hydrogen ions between the solution and the sorbent is a relatively slow process) and b) a static one, (multipoint protein fixation due to the interaction of enzyme and carrier charged groups).

Changes in the enzyme stability after its covalent binding to soluble or insoluble carriers depend on several factors: 1) the chemical modification of enzyme functional groups, 2) a polyfunctional, multipoint interaction between an enzyme and a carrier, 3) changes in the enzyme microenvironment. Usually, these factors work simultaneously and sometimes act in opposite directions. So, to choose an optimal immobilization method for the given enzyme, one should be certain of the relative significance of each of these factors.

The chemical modification of enzymes with low molecular weight monofunctional agents, in order to get a stabilized enzyme derivative, is frequently a natural event. It is already known that single mutations in aminoacid composition of enzymes can cause thermostability changes of the enzyme. The comparison of aminoacid composition and thermostability of some amylases and proteases, purified from different strains of mesophilic and thermophilic microorganisms, makes it possible to define some structural features influencing protein thermostability [42].

Chemical modification of an enzyme is also an indispensable event during covalent enzyme immobilization on a polymeric carrier. The investigation of enzyme stability as a function of the number of modified functional groups opens the opportunity to classify different functional groups, according to their role in the maintenance of the enzyme native structure [43].

Bifunctional cross-linking agents are often used to obtain intramolecularly cross-linked enzymes or soluble or insoluble enzyme aggregates. The introduction of intramolecular cross-links into an enzyme globule makes the protein molecule more rigid and less capable of unfolding.

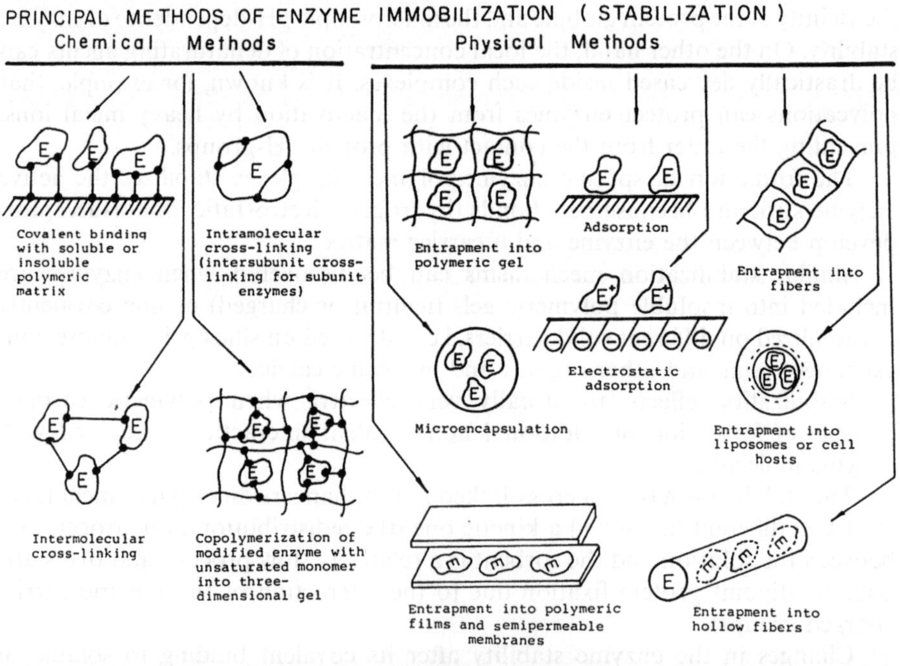

Fig. 3. Principal methods of enzyme immobilization (stabilization)

The modification of enzymes by water-soluble polymers or other proteins is usually used to obtain stabilized and still soluble enzyme derivatives.

There exist methods to incorporate enzymes into polymeric membranes with subsequent protein fixation in the membrane by glutaric aldehyde. Protein membranes (mainly, made of albumin) with covalently linked enzymes have also been described. Upon immobilization into membranes, many enzymes preserve 30–80% of their initial activity and demonstrate very high thermostability. The stabilization effect in this particular case is caused mainly by:

a) the fixation active conformation of the protein and
b) the high local concentration of stabilizing proteins. One can assume that mixed enzyme-protein membrane resemble the structures in which enzymes are active in nature.

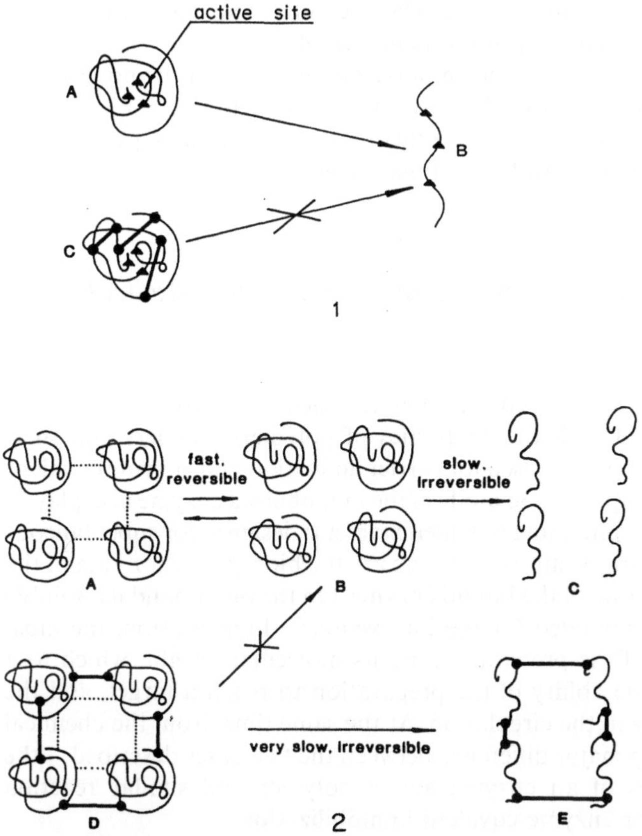

Fig. 4. Mechanism of enzyme active conformation stabilization by intramolecular cross-linking with bifunctional reagents (**1**) single-chain enzyme; *A*—native enzyme, *B*—denaturated unfolded protein, *C*—intramolecularly cross-linked stabilized enzyme; (**2**) subunit enzyme: *A*—active enzyme, *B*—reversibly dissociated subunit enzyme, *C*—irreversibly unfolded subunits, *D*—intramolecularly intersubunit cross-linking of subunit enzyme, *E*—irreversibly denaturated cross-linked subunit enzyme. (Reproduced with permission from Torchilin VP (1987) Adv Drug Deliv Rev 1:41

Often, enzymes are included into artificial microcapsules. Inside these carriers an enzyme is protected from many agressive influences of the external medium and, in turn, does not act on different external objects. This approach includes also enzyme immobilization by entrapment into artificial phospholipid vesicles—liposomes or cells and their ghosts.

Surely, not all of the approaches listed are of practical interest for obtaining immobilized forms of therapeutic enzymes. It is evident that the most promising methods are connected with the limitation of the conformational mobility of the protein globule and with the spatial separation of the enzyme molecules and natural endogenous inactivators. Into the first group one can include such methods as chemical modification; intra- and intermolecular cross-linking with bifunctional reagents; entrapment into polymeric gels, fibers and films; multipoint covalent immobilization on different carriers; enzyme copolymerization with gel-forming monomers. The second group includes enzyme entrapment into microcapsules, liposomes and cell ghosts.

Some principal methods of enzyme immobilization by chemical and physical methods and their stabilization are schematically presented in Fig. 3. Fig. 4 gives more details about the stabilization of single-chain and subunit proteins by intramolecular cross-linking with bifunctional reagents.

3.2 The Peculiarities of Enzyme Immobilization for Medical Use. Carriers and Immobilization Methods Required

In the preparation of immobilized enzymes for medical purposes, the main limitations are connected with the properties of polymeric carriers and their intermediate products which are used for carrier activation and enzyme covalent binding. The subsequent application mode of the immobilized enzyme also plays a role. Thus, if the enzyme-containing polymeric tablet is designed for subcutaneous implantation, the most important parameters are the biodegradation rate of the polymer and the release rate of the bound enzyme. On the other hand if a soluble immobilized enzyme is intended for the intravenous administration, the most important parameters of the preparation are its molecular weight, which to a great extent regulates the ability of the preparation to reach a target, and the time of drug availability in the circulation. At the same time, from the chemical point of view, there is no major difference between the two cases described—the same functional groups of an enzyme and a polymer and similar reaction conditions were used for enzyme covalent immobilization.

Both natural and synthetic high-molecular-weight compounds of different structure are used for therapeutic enzymes modification and immobilization [45]. Enzymes are usually immobilized by one of the following ways:

1) by mechanical entrapment (e.g., by microencapsulation). A particular case of this method is enzyme non-covalent incorporation into a polymeric gel,

2) by direct intermolecular interaction, including the formation of electrostatic or chemical bonds between functional groups of a carrier and that of an enzyme to be immobilized,
3) by polymerization and copolymerization or polycondensation and copolycondensation of corresponding functional derivatives of an enzyme and a carrier,
4) by the treatment of an enzyme with bifunctional or polyfunctional cross-linking agents,
5) by covalent binding of an enzyme with polymeric carriers via intermediate cross-linking agents.

Despite of the variety of chemical reactions used in immobilization processes, the major part of immobilized enzymes in medical application is obtained utilizing a relatively narrow circle of natural and synthetic polymeric carriers. These polymers contain functional groups which can participate in immobilization reactions directly or upon preliminary activation.

Polymeric carriers and modificators intended for medical use should, evidently, meet the following requirements:

1) the complete absence of toxicity of both the polymeric carrier and the product obtained upon immobilization,
2) the ability to clear from the organism via metabolic transformations or by direct excretion; in the latter case, the molecular weight of the carrier should be controlled and the carrier should be completely free of high molecular weight fractions incapable of kidney filtration,
3) the presence of reactive groups capable of chemical interaction with an enzyme and incapable of side-reactions (hydrolysis, cross-linking, etc.),
4) the ability not to cause, under immobilization conditions, any chemical or physico-chemical changes in the properties of the enzyme to be immobilized, which could affect its normal biological activity (side-modification, denaturation),
5) high degree of purity (absence of traces of monomers, solvents or toxic low molecular weight fractions).

The main types of polymers, which are usually used for therapeutic enzymes immobilization and modification are reviewed in [46–49]. Here, we will briefly discuss only the most important properties of these carriers.

Natural polymers. Natural polymers (polysaccharides, proteins) are easily available and find wide application. They are hydrophilic, which permits all manipulations with these carriers in water.

The general drawbacks of these carriers and modifiers are their sensitivity to microorganisms and difficulties in obtaining polysaccharides in technological shapes such as tubes, spheric granules, tablets, and porous structures (unlike synthetic polymers).

The most often used polysaccharide carriers are derivatives of cellulose, starch, agarose and dextran. Cellulose or 1,4-β-D-glucopyranosyl-D-glucopyranose (sometimes containing defective units with 1,6-bonds) can be used as carrier. Its derivatives (with known degree of substitution) can be used for protein binding directly, (e.g., bromoacetyl cellulose), or after preliminary activation, (e.g., amido- or amino-derivatives, which can be transformed into reactive hydrazido- or diazo-derivatives).

Agarose or poly-β-galactopyranosyl-3,6-anhydro-α-L-galactopyranose is also often used as a carrier for immobilization. An agarose water solution when cooling below 45 °C forms firm gels, which can be additionally cross-linked with bifunctional reagents.

Many carriers are obtained from dextran-branched or linear polysaccharide containing glucose residues, linked mainly with 1,6-bonds. Dextran gels are chemically stable. It is important that water-soluble dextran derivatives containing different reactive groups which easily metabolize can be prepared. These properties make dextran especially interesting as a drug carrier.

Chitin and its deacetylated derivative chitosane are among other potentially useful polysaccharide carriers for enzyme immobilization.

Proteins and their derivatives are relatively rarely used as carriers, despite of the data existing on enzyme immobilization on albumin, collagen and fibrin.

Synthetic polymers. Many synthetic polymers are used for enzyme (and other biologically active compounds) immobilization and modification. The main advantage of synthetic polymers is the easy in varying their physical properties, chemical structure, nature of reactive groups, and micro- and macro-environment.

Polyacrylamide as a water soluble homopolymer can be a good starting substance to obtain hydrophilic carriers for protein immobilization. Granulated polyacrylamide gels are often used which can be additionally derivatized with different reactive groups. Enzymes and cells or cellular organelles can be entrapped into polyacrylamide gel in the process of acrylamide copolymerization with a cross-linking reagent like methylene-bis-acrylamide in water.

Spherical granules can be made from copolymers of 2-hydroxyethylmethacrylate and ethylenedimethacrylate. These copolymers are highly hydrophilic, biologically inert and are not attacked by microorganisms. The required reactive groups can be easily introduced in these carriers by conventional chemical methods.

Different granules, fibers, tubes, films and membranes are usually prepared from polyamides. The surface of polyamide-made products can be chemically activated by partial hydrolysis of surface amide groups with the formation of amino and carboxy groups.

Water-soluble polymers also play an important role in the modification and immobilization of therapeutic enzymes. Many of them are produced on an industrial scale.

Unfortunately, in many cases, polymers do not clear from the organism and may provoke serious complications. The data available at present permit to assume the existence of two different mechanisms, which determine "post-effects" of water-soluble polymers in the body. The first is connected with kidney filtration (rate constant K_1), the second—with polymer capture by the cells of the reticulo-endothelial system (rate constant K_2). During the process of membrane ultrafiltration in the kidneys, polymers with a globule size less than 2.5 nm are cleared from the body with the maximal rate K_1; macromolecules having the size of about 3.5 nm are cleared with the maximal rate $0.5\,K_1$; and transmembrane clearance drops to zero if the size of macromolecules in solution is 6.0 nm and more [47].

It was shown that with the increase in M_W/M_N ratio (i.e., the increase in polymer heterogeneity) the toxicity of polymers also increases. It is probably connected with the presence of both low-molecular-weight and high-molecular weight fractions, which are usually more toxic than fractions of intermediate molecular weight [45].

The ability of high molecular weight polymers to accumulate in the organism leads to the development of new synthetic and semisynthetic materials capable of transformation into non-toxic products under physiological conditions.

The binding chemistry is quite similar for both soluble and insoluble immobilized enzyme preparation. Nevertheless, in each case specific problems can be encountered. Thus, in the use of insoluble enzyme-containing polymeric particles or implants, the main difficulties are connected with carrier biodegradation in physiological conditions and the continuous and uniform release of an immobilized enzyme. At the same time, the ability to reach extravascular targets is questionable after intravenous administration of soluble immobilized enzyme preparations. The first barrier for a drug after intravascular administration is to cross the layer of endothelial cells; the penetration through this layer often becomes the limiting step in the total process of drug distribution over the whole organism. It was shown [50] based on numerous published data that enzyme:polymer conjugates can pass through the endothelium almost exclusively via transcytosis mechanisms, which is a rather rare event (at least for a carrier-bound enzyme), see scheme in Fig. 5.

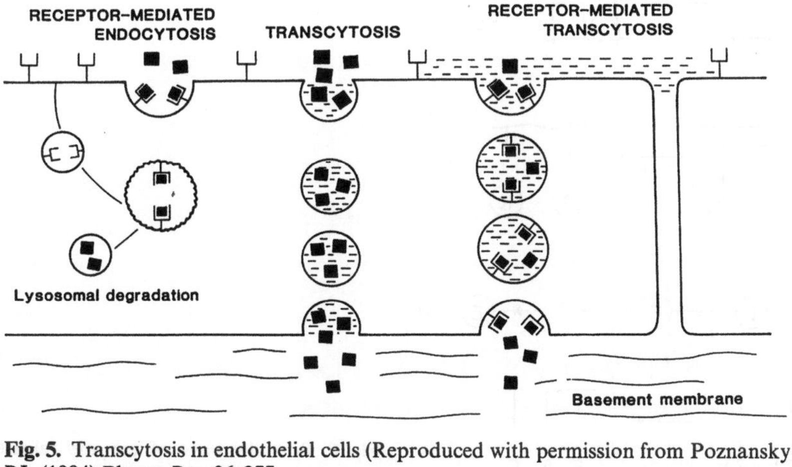

Fig. 5. Transcytosis in endothelial cells (Reproduced with permission from Poznansky MJ, Juliano RL (1984) Pharm Rev 36:277

Thus, intravascular administration of immobilized enzymes usually permits to reach only intravascular targets. The very important barrier to overcome is the reticulo-endothelial system, which is specially designed by evolution to remove foreign particles, aggregates and macromolecules from the circulation. Both barriers can be crossed by an enzyme, which is reversibly immobilized on a soluble or insoluble polymeric matrix and can be released slowly from this matrix (possible ways to obtain preparations capable of free enzyme release are shown schematically in Fig. 6).

The main approaches in the use of soluble and insoluble polymers for drug immobilization are considered in [46, 47]. It follows from the data reviewed that the use of a polymeric carrier, demonstrating specific, uneven organ distribution, permits also to influence the distribution of the immobilized drug to some extent. In [51, 52] polyhydroxy-propylmethacrylamide, polyvinylpyrrolidone and some other polymers are referred to as the most promising means for drug delivery. Chemical aspects of the synthesis of macromolecules containing physiologically active groups, by preliminary modification of the corresponding bioactive compounds with some kind of unsaturated modificator are discussed in [53]. This leads to the formation of a so-called macromonomer, which is than copolymerized with a polymerizable low molecular weight monomer, (e.g., acrylamide), giving a three dimensional polymeric structure. Soluble dextran is still very attractive for biomedical application [54], mainly due to its biocompatibility and ability to be degraded in the body. Insoluble polymers, (such as cross-linked polymers and copolymers of vinylic series) designated for the use in extracorporeal devices and implants, also have been frequently studied. It was suggested to use polymethylmetacrylate, modified with different biologically active compounds, in extracorporeal hemoperfusion systems [54].

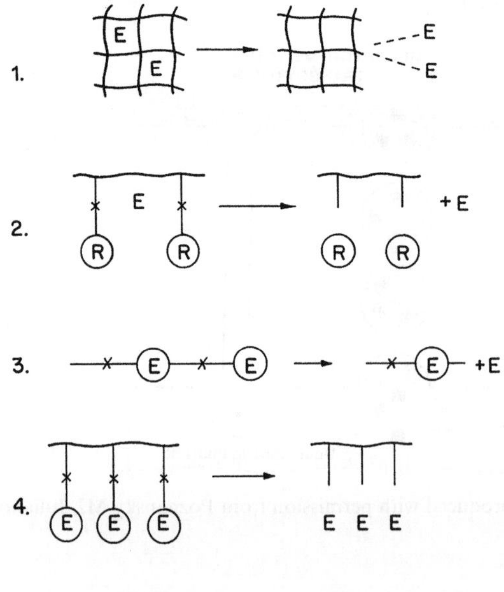

Fig. 6. Possible ways of enzyme release from an immobilized preparation

During the last 10 years, the synthesis and biomedical use of synthetic polymers, containing in the main chain or in side-chains chemical bonds capable of destruction by endogenous enzymes, was studied extensively. Besides different polyaminoacids, polymers consisting of relatively long, but still excretable blocks (molecular weight approx. 40 000), connected with splittable peptide or peptide-like bonds [49] are being investigated. Biodegradable polymers represent the future of polymeric enzyme derivatives [55]; they have also been used for the synthesis of preparations, capable of specific accumulation in definite compartments of the body [56–61]. Data are available on the use of biodegradable polyanhydrides for immobilization of biologically active compounds [62]. Among bio-degradable polymers which solubilize in physiological media and can be further "cut" into small and easily excretable fragments, one should mention polylactic acid, polyglycolic acid, polyesters, polycarbonates, ethyleneoxide-ethyleneterephtalate copolymers, polyortho-esters, polyacetals, polyorthocarbonates, copolymers of oxalates with linear and cyclic diols, polyphosphoric esters, poly-alfa-cyanoacrylates, poly-ε-capro-lactam, poly-alfa-amino acids and some others (several examples are presented in Fig. 7).

The immobilization process is based on the interaction between reactive groups of a polymeric carrier and an enzyme. Principally, different functional groups of amino acids residues can participate in the reaction: carboxylic groups of glutamic and aspartic acids and terminal carboxy-groups; aminogroups of lysine and hydroxylysine and terminal aminogroups; phenolic groups of tyrosine; imidazole groups of histidine; indol groups of tryptophane; SH-groups of cysteine; disulfide groups of cystine; hydroxy-groups of serine, hydroxyproline, hydroxylysine and hydroxyglutamic acid; guanidine groups of arginine;

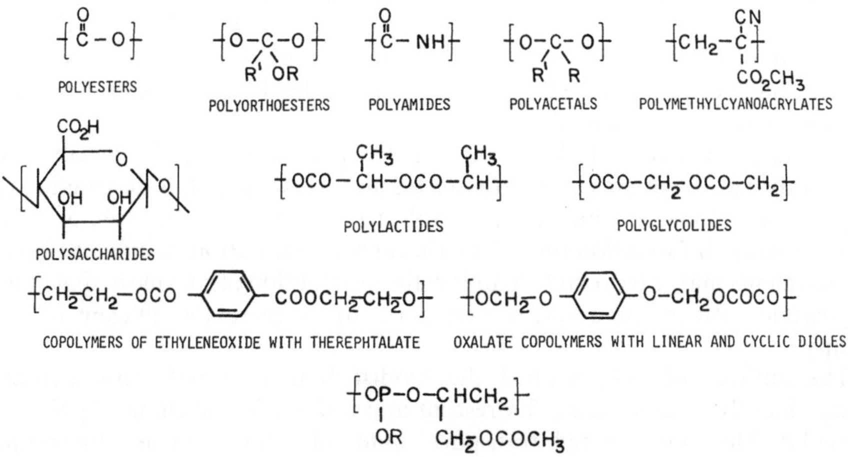

Fig. 7. Biodegradable polymers (as repeated units) (Reproduced with permission from Torchilin VP (1987) Adv Drug Deliv Rev 1:41

methylmercapto-groups of methionine; phenyl groups of tyrosine and phenylalanine; amidogroups of asparagine and glutamine; and in some special cases, even peptide groups of the main chain [32–38, 45].

Among binding processes, the most frequent are substitution and addition reactions, proceeding at aliphatic and aromatic carbon atoms and heteroatoms. Radical processes are relatively rare.

Without going into the chemistry of the reactions mentioned, which can be found in [33, 37, 45], we will just enumerate the most important methods of enzyme binding with reactive carriers. The covalent methods are: enzyme acylation with polymeric acyl-azides; acylation in the presence of carbodiimides; acylation in the presence of Woodward's reagent; acylation with polyanhydrides; acylation with cyclic carbonates of polyoles; azocoupling; the formation of azomethine derivatives; the use of polyimidocarbonates, polymers with isocyanate and isothiocyanate groups or active halogen atoms (chlorotriazine derivatives or derivatives with active halogens at aliphatic or aromatic carbon atoms or with active halogen atoms in inorganic groups); the use of polymeric imidoesters and carriers with activated double bonds; the use of different radical and ionic reactions. We do not mention here multiple methods of mechanical entrappment of enzymes into polymeric gels or microencapsulation.

3.3 *The Fate of Immobilized Enzymes in the Body*

The fate of an immobilized enzyme in the organism depends to a great extent on the fate of the polymeric carrier used for immobilization. What can be said about the expected behaviour of an immobilized enzyme based on the required properties of an enzyme:polymer conjugate (see schematic presentation, taken from [63] in Fig. 8) and the chemistry of a polymeric carrier? First of all, in each separate case it is necessary to know the possible reactions of surrounding tissues, blood and other biological fluids on the foreign polymeric materials. This problem has been considered in detail [64]. Evidently, all undesirable side-reactions have to be excluded.

Second, it is necessary to know the routes of polymeric material distribution in the body, i.e., the organs where it is accumulated, the rate of its destruction (if the polymer is biodegradable), and finally the route and the rate of its elimination from the body. Information on polymeric carrier distribution in different body compartments makes it possible to follow the distribution of enzyme activity and its possible increase in a target organ, i.e. to estimate the efficacy of the therapy.

The authors of [47] studied the biodistribution of polymeric carriers (predominantly soluble) using fluorescent and radioactive labelling of polymer molecules. They investigated transport routes of soluble carriers (including penetration through the endothelium, transport into the lymphatic system, possibility of glomerular filtration and deposition of polymers in intracellular compartments) and the influence of the polymer chemical structure, properties

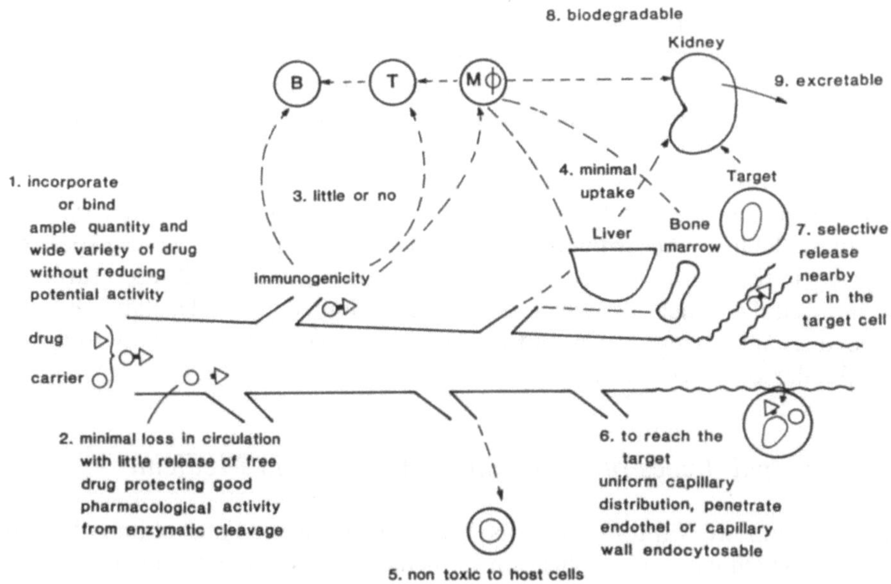

Fig. 8. Schematic illustration of requirements for drug carriers (Reproduced with permission from Suzuta T (1983) In: Bruck SD (ed) Controlled drug delivery, CRC Press, Boca Raton, vol 2, p 163

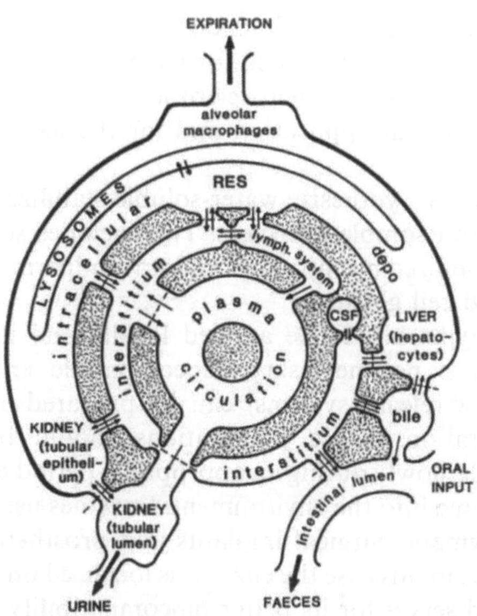

Fig. 9. Multicompartment model of organism. Arrows depict corridors connecting body compartments and represent possibilities of polymer transfer between compartments either restricted by compartmental barriers (*dashed lines*) or occurring as flux transfer (*simple arrows*) (Reproduced with permission from Drobnik J, Rypacek F (1984) Adv Polym Sci 57:1

and administration method on its biodistribution and fate in the organism. They suggested a multicompartmental model of the organism (Fig. 9) and showed that in the process of migration, polymers can participate in reactions of modification, hydrolysis and oxidation and in the immune response. An important conclusion made by the authors is that the blood plasma serves as the medium delivering polymers to different compartments of the organism.

Polymers are cleared from the body via the reticulo-endothelial system, mainly by the liver. According to the model suggested in [47], the transport of polymers in the liver proceeds in several ways. Among them is diffusion through the intercellular junctions, which is a process limited by molecular size, and pinocytosis into hepatocytes and Kupffer cells, which is the main route of intracellular polymer delivery.

3.4 *Principal Types of Immobilized Enzymes in Medicine*

In this section, we will try to describe briefly the principal types of immobilized enzymes, which can find application in medicine as therapeutic agents.

Immobilized enzymes as drugs can be classified in two different ways. The first is based on their distribution according to targets of potential therapy (see Table 4). It is easily to notice that the table contains data on native enzymes, which are already in use for treatment of certain diseases and on possible methods of their immobilization and stabilization. These data have been discussed in Sects. 2.1 and 3.1.

The second classification is based on possible administration methods of immobilized enzymes used in therapy. This classification will be the base for data presentation in this book. From this point of view, the following groups of preparations can be named: 1) immobilized enzymes intended for prolonged circulation in the blood, 2) immobilized enzymes intended for local administration in the therapy of injuries like thrombi, tumors, atherosclerotic deposits etc., 3) immobilized enzymes intended for external application and for the use in extracorporeal systems.

In the first group, it seems reasonable to synthesize water-soluble stabilized enzymes by modification of enzymes with soluble polymers. This includes so-called "artificial cells", i.e. enzymes immobilized by entrappment into semipermeable microcapsules, liposomes, cells and cell ghosts.

In the second group, when an enzyme has to be applied locally and its presence in other organs and tissues is not necessary, biocompatible and biodegradable enzyme derivatives (slow-release systems) can be prepared as microparticles or tablets for implantation. These preparations should be localized in the affected zone and degrade slowly during an appropriate period of time, continuously releasing active enzyme into the environment. Enzymes used for modification of the surface of polymeric surgical implants and prosthetic devices also belong to the group, because in this case the enzyme is localized only in the prosthetic device or implant and serves for its better biocompatibility.

Table 4. The principal types of immobilized therapeutic enzymes

Preparation, possible methods of application	Typical examples	Type of action	Possible methods of immobilization
Enzymes for the replacement therapy in the treatment of digestive organ diseases, (implantation, parenteral application)	pepsin, chymotrypsin, trypsin, amylase, lipase or their mixtures	local	microcapsulation, entrapment into polymeric gels and particles
Antitumor enzymes (implantation; intravascular, parenteral or extracorporeal application)	asparaginase, arginase, nuclease, desoxyribonuclease	local and/or systemic	microencapsulation, entrapment into liposomes, entrapment into red blood cells ghosts, entrapment into gel particles, immobilization on soluble polymers
Enzymes for the therapy of inherited lysosomal enzymes insufficiency (intravascular application)	glucosidase, glucuronidase, galactosidase	local	entrapment into liposomes (with facultative preliminary stabilization by intermolecular cross-linking or immobilization on soluble polymers)
Enzymes for the thrombolytic therapy (intravascular application)	plasmin, urokinase, streptokinase, tissue plasminogen activator	local and/or systemic	immobilization on soluble polymers, entrapment into red blood cells ghosts, immobilization on biodegradable microparticles
Antibacterial, antiviral, antiallergic enzymes (intramuscular or intravascular application)	penicillinase, lysozyme, nucleases	systemic	immobilization on soluble polymers, entrapment into microcapsules, red blood cells ghosts, liposomes
Hydrolytic enzymes for the action on pathological tissues and anti-inflammatory enzymes (local application)	trypsin, chymotrypsin, collagenase, papain, ribonuclease, desoxyribonuclease, peptidase, lysozyme	local	immobilization on materials for dressing, drainages, tampons (films, fibres, etc.), entrapment into microcapsules, immobilization into powdered polymeric gels

The use of immobilized enzymes for external application and in extracorporeal devices opens wide possibilities. In this case, an enzyme can serve as a component of biological reactors (being immobilized in the structure of granules, fibers, films, microcapsules, etc.); it can be used for the modification of the inner surface of different extracorporeal devices in order to increase their compatibility will blood; it can be immobilized in fibers and films used for the manufacturing of drainages and dressings and incorporated into powders and ointments, used in the treatment of wounds, ulcers and burns.

In this order we will consider the present state of immobilized enzymes use in medicine.

4 Therapeutic Immobilized Enzymes for Parenteral Application

4.1 *Preparations Intended for Prolonged Circulation in the Body*

As stated above, therapeutic immobilized enzymes capable of prolonged functioning in the body (in the blood) without loss of specific activity can be obtained by different methods [65–67]. The most popular of them are enzyme immobilization on soluble polymeric carriers or their entrapment into "artificial cells" (microcapsules, liposomes, cells and cell ghosts). These preparations are also supposed to affect the immune system of the organism to a lesser extent than their native precursors. At present, it is difficult to judge which of two approaches mentioned is more promising. Nevertheless, historically the majority of the early studies dealt with enzyme modification using soluble polymers.

4.1.1 *Enzymes Modified with Soluble polymers*

4.1.1.1 Carriers and Methods of Immobilization; Properties of Immobilized Enzymes

From the medical point of view, a carrier polymer for the preparation of parenteral drugs should possess the following properties: high purity and homogenicity in respect of both composition and molecular-weight distribution; complete biocompatibility, i.e. the absence of any undesirable reactions in the site of application and in the whole organism; the ability to undergo biodegradation with a reasonable rate, and the absence of any biological activity in the products of biodegradation; and easy clearance from the body after the function of a polymer-bound drug has been fulfilled.

A great deal of attention has been paid to obtain synthetic polymer carriers, as there exist considerable possibilities for regulating the properties of both the carrier itself and a bound enzyme. A detailed review on synthetic carriers intended for medical enzyme immobilization is given in (Refs. [45–47, 64]) (see also Fig. 7).

Naturally, the examples presented do not exhaust the number of polymers used. The use of polyurethanes, polycarbonates, polyepoxides and some other polymers as carriers for medical enzymes has been reported [68, 69].

Furthermore, some success has been achieved in the synthesis of polymeric carriers capable of selective accumulation in biological targets [60, 61, 70].

It is very important to know the behaviour of a (bio)polymer after its administration. An inert chemical polymer may cause undesirable side-effects in the body if its molecular weight exceeds some critical value. Thus, biologically stable polymers, or fragments of biodegradable polymers, which should presumably be released from the body via the kidneys, may accumulate there, decreasing the efficiency of the kidney filtration if a polymer molecular weight exceeds 60–80 kDa. This phenomenon may be termed as "physical toxicity". Naturally, the situation is seemingly more simplified for biodegradable polymers. There exists a number of approaches for the synthesis of polymers capable of splitting under physiological conditions [45, 46]. As enzymatic degradation is the main route of biodegradation, the introduction of groups which resemble natural substrates of lytic enzymes (peptide, saccharide or nucleotide bonds) is the prime approach to obtain biologically degradable polymers.

What are the main effects achieved after the multipoint binding of enzymes (and other proteins) with water-soluble polymers? First, there is an increase in conformational stability and stability towards the action of endogenous proteases, due to the presence of a carrier molecule and the chemical modification of lysine residues sensitive to trypsin-like proteases by moieties which create steric hindrances for hydrolysis [70]; second, there is an increase in the circulation half-life of the active enzyme, due to increase in its stability and decrease in the rate of its filtration by the kidneys, as well as by the prevention of any interaction with cell receptors, which could result in removal of the proteins from the circulation; third, protein modification by polymers enables a direct regulation of the host's/immune response.

In early studies, natural soluble polysaccharides (such as dextran) were the most popular carriers for enzyme immobilization. To achieve enzyme-to-carrier covalent binding, dextran was activated with cyanogen bromide [71–75]. It has been shown that the immobilization on BrCN-activated dextran can drastically increase the thermostability of α-amylase, β-amylase, trypsin and catalase in *in vitro* conditions (see Figs. 10–13). Also, the polymeric matrix protects the

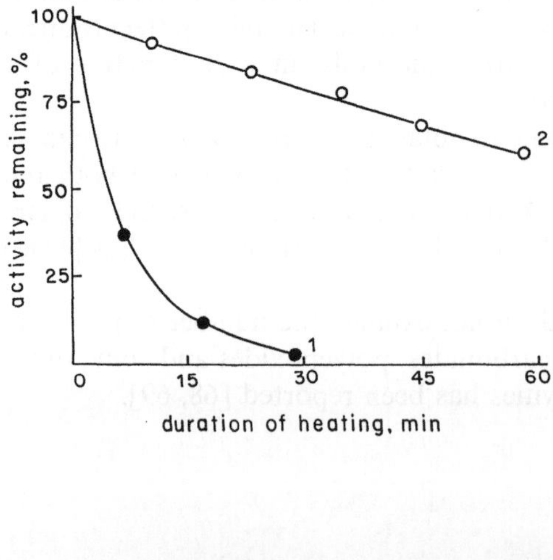

Fig. 10. Loss of activity after heat treatment at 65 °C of *Bacillus amyloliquefaciens* α-amylase (●) and its conjugate with dextran (○). Native and modified enzyme solutions (5.5 μg of protein/ml) in water were heated and samples removed for activity determination at different times. α-Amylase activity was determined by the release of reducing sugars from soluble starch by using the dinitrosalicylic acid method (Reproduced with permission from Marshall JJ, Rabinowitz ML (1975) Arch Biochem Biophys 167:777

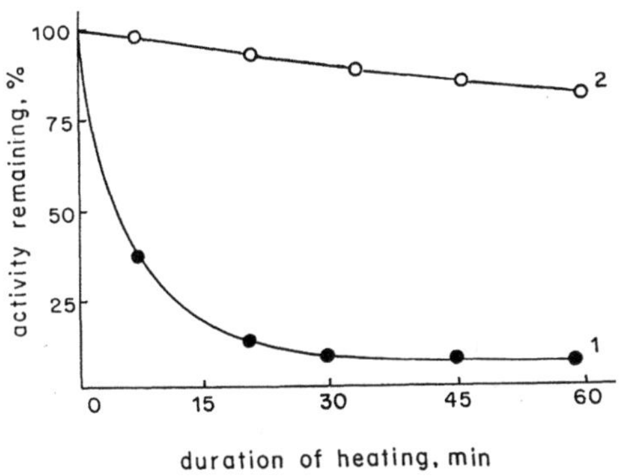

Fig. 11. Loss of activity after heat treatment at 60 °C of sweet-potatoe β-amylase (●) and its conjugate with dextran (○). Samples (50 μl) of native and modified enzyme (activity 0.5 U/ml) were heated for different lengths of time and the activity remaining determined by a copper-reducing power method measuring the release of reducing sugars from starch (Reproduced with permission from Marshall JJ, Rabinowitz ML (1975) Arch Biochem Biophys 167:777)

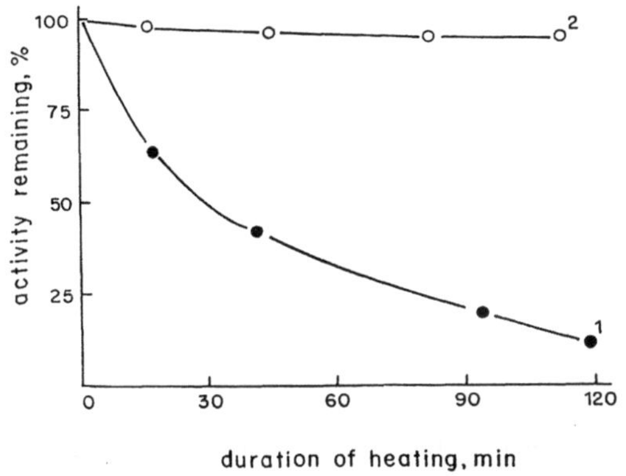

F'g. 12. Loss of activity after heating ypsin (●) and its conjugate with dextran (○) at 37 °C. Samples (125 μl) of native and modified enzyme (activity 2 U/ml determined against *p*-toluene sulfonyl-L-arginininemethyl ester) were heated for different lengths of time in 80 mM borate buffer pH 8.1; the remaining activity was determined by the method of Hummel (Reproduced with permission from Marshall JJ, Rabinowitz ML (1975) Arch Biochim Biophys 167:777)

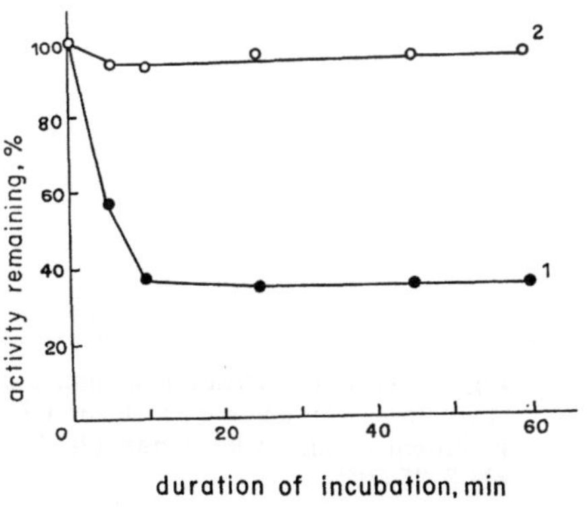

Fig. 13. Heat inactivation of native catalase (●) and catalase-dextran conjugate (○). The 2 forms of the enzyme (8 μg protein/ml) were held at 52 °C and samples were removed for assay at stated intervals (Reproduced with permission from Marshall JJ et al. (1976) Biotechnol Bioeng 18:1325)

immobilized enzyme from the action of proteases (Fig. 14) and decreases the autolysis rate for immobilized proteolytic enzymes (Fig. 15).

Other activation methods for polysaccharide carriers have been also suggested in order to exclude highly toxic cyanogen bromide from the procedure. One of them is the incorporation of free aldehyde groups into polysaccharides [76, 77] by oxidation of glucoside units with periodates. The aldehyde groups of a carrier easily interact with terminal and ε-amino groups of lysine residues resulting in Schiff bases formation. Schiff bases, which slowly hydrolyze under physiological conditions, can be reduced by sodium borohydride or cyanoborohydride to stable bonds. Simultaneously, the reduction of non-reacted aldehyde groups into inert hydroxy groups also occurs.

Pronounced stabilization effects can be achieved by multipoint immobilization of enzymes on vinylic polymers and copolymers. Polyvinylpyrrolidone (biocompatible plasma extender and detoxifying agent) is often used as a carrier. In order to make it capable of protein chemical binding, vinylpyrrolidone is copolymerized with a reactive comonomer (e.g., acrylic acid).

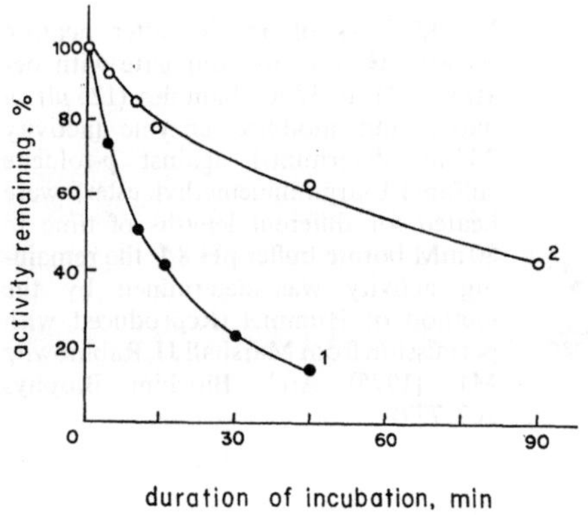

Fig. 14. Inactivation of native catalase (●) and catalase-dextran conjugate (○) by trypsin (Reproduced with permission from Marshall JJ et al. (1976) Biotechnol Bioeng 18:1325)

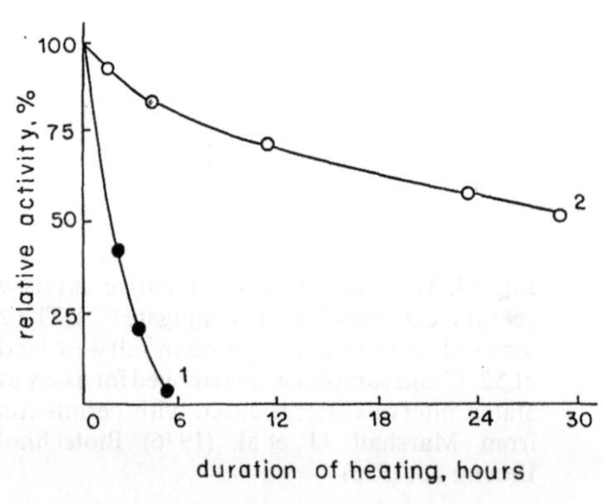

Fig. 15. Thermoinactivation of native α-chymotrypsin (*1*) and α-chymotrypsin, immobilized on aldehyde-dextran, (*2*) 37 °C, pH 8, 25 (76)

As a result, a protein can be bound with the carrier via carbodiimide activation of carboxylic groups of acrylic acid units (see Fig. 16).

The natural result of the increased stability of enzymes immobilized on soluble carriers is the increase in their half-life in the circulation of experimental animals after intravenous administration. The increase in the half-life of an enzyme usually correlates well with the increase in the stability of this enzyme in vitro. The data illustrating this observation are presented in Figs. 17 and 18.

In one of the initial studies, it was shown that α-amylase and catalase bound with BrCN-activated dextran have a slow clearance from the blood in rats [78], so that 2 h after intravenous administration, 75% of immobilized α-amylase (but only 16% of the native enzyme) remained in the circulation. After 2 h, the activity of immobilized catalase was higher than 50% of the initial blood levels, whereas the native enzyme had disappeared from the blood within 1 h. It was also shown that after intraperitoneal administration of immobilized enzymes, their activity in the blood was found over a more prolonged period of time: i.e., native catalase was not found 4.5 h after administration, whereas the immobilized enzyme

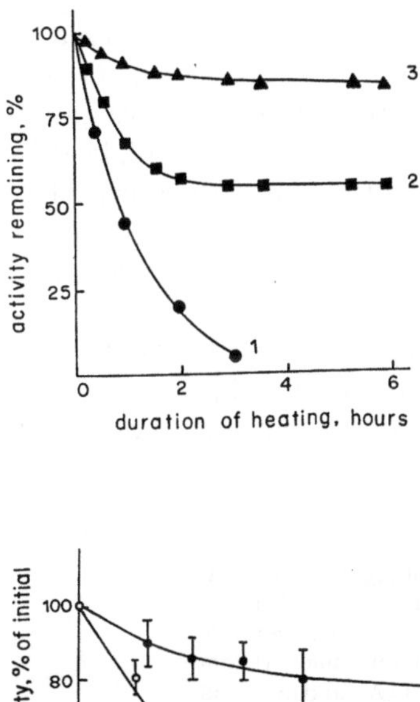

Fig. 16. Thermoinactivation of native α-chymotrypsin (*1*) and α-chymotrypsin immobilized on vinylpyrrolidone-acrylic acid copolymer: (*2*)—acrylic acid content 1% mol, (*3*)—acrylic acid content 10% mol. The increase in acrylic acid units concentration leads to the increase in immobilized enzyme thermostability due to formation of multiple enzyme-polymer bonds

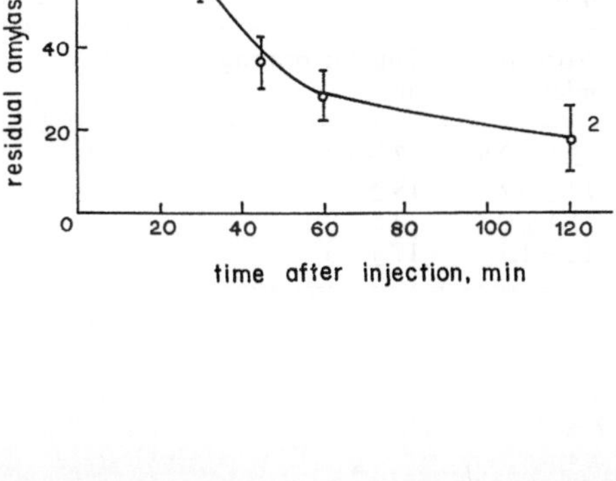

Fig. 17. Removal of *Bacillus amyloliquefaciens* α-amylase from the circulation of rats after intravenous administration of native enzyme (○) or α-amylase-dextran conjugate (●). In both cases 400 U amylase activity (1 mg protein in the case of native amylase and 2,5 mg protein in the case of conjugated amylase) were injected. The average values and ranges for groups of three rats are shown (Reproduced with permission from Marshall JJ et al. (1977) FEBS Lett 83:249)

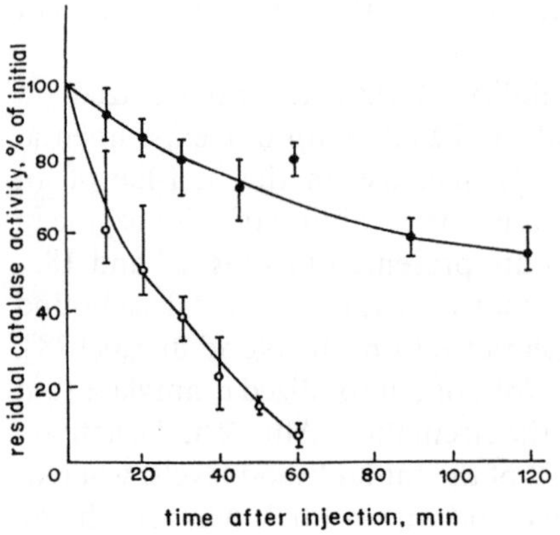

Fig. 18. Removal of bovine liver catalase from the circulation of acatalasemic mice after intravenous administration of native enzyme (○) or catalase-dextran conjugate (●). In both cases, 12000 U catalase (approx. 1 mg protein) were injected. The average values and ranges for groups of four mice are shown (Reproduced with permission from Marshall JJ et al. (1977) FEBS Lett 83:249)

exhibited more than 30% of its initial activity 22 h after administration. It has been also shown that the use of immobilized amylase decreases a host immune response [79], decreasing the ability of immobilized enzyme to interact with antibodies against the native enzyme (the primary response remained unchanged).

Analogous results have been obtained with immobilized enzymes used as antitumor agents. Thus, the covalent binding of carboxypeptidase G and arginase with soluble BrCN-treated dextran—with a molecular weight of 40000—considerably prolongs the enzyme half-life in the blood of both healthy mice and mice with inoculated tumors [80]. The half-life for carboxypeptidase clearance, 3.5 h for normal mice and 7 h for mice with tumors, is increased up to 17–18 h for dextran-bound enzymes. The half-life for arginase clearance is also increased, from 1.4–2.5 to 12–17 h (see Table 5).

Using subtilysine, the reaction conditions were investigated permitting to vary the enzyme-to-carrier ratio in the resulting conjugate [81]. It was shown that variations in this ratio can noticeably influence the stability and the catalytic activity of the immobilized enzyme.

Table 5. Persistence of native enzyme and dextran-enzyme conjugates in normal and tumour-bearing DBA/2 mice. Half-lives (t_f) of the enzymes were calculated from semi-logarithmic plots of enzyme activity in the blood against time, and are shown as means ±S.D. of results from groups of three of four mice. (From Sherwood R.F., Baid J.K., Atkinson T., Wiblin C.N., Rutter D.A., and Ellwood D.C., Biochem J., 164, 461, 1977, with permission)

	t_f(h)	
Enzyme	Normal mice	Tumour-bearing mice
Native carboxypeptidase G	3.5 ± 0.5	7 ± 1.2
Dextran-carboxypeptidase G conjugate	17 ± 3.2	18 ± 2.0
Native arginase	1.4 ± 0.2	2.5 ± 0.3
Dextran-arginase conjugate	12 ± 1.7	17 ± 2.5

A comparative analysis of preparations obtained by β-galactosidase (the enzyme used for the treatment of genetic storage diseases) modification with dextran, acetylated dextran and methylated dextran has been reported [82]. The authors have shown that the modification increases enzyme stability towards thermoinactivation and action of proteolytic enzymes. Higher stability can be achieved with hydrophilic polysaccharide carriers—dextran and acetylated dextran, than with hydrophobic methylated dextran.

The therapeutic activity of hyaluronidase, immobilized on aldehydedextranes of different molecular weights, in the treatment of experimental silicosis has been studied in rat experiments [83, 84]. The immobilized enzyme was administered into rats by inhalation or intraperitoneally during 4 months. It has been shown that the inhibiting action of the immobilized enzyme on the development of fibrosis in the lung was much more pronounced, than that of the native enzyme.

The protection afforded to enzymes by linking them to biocompatible synthetic polymers and copolymers was described by Abuchowski et al. [85], who showed in mice that catalase immobilized on different polyethyleneglycols (via 2,4,4-trichlorotriazine) retains 90% of its initial activity and is removed from the circulation several orders of magnitude slower than the native enzyme. In the same way, the covalent binding of urease and trypsin with polyethyleneglycol noticeably increases the circulation time of enzymes in the blood of experimental animals [86, 87]. Also, this group has shown that modification of glutamyl-L-asparaginase, phenylalanine ammonia lyase and uricase of animal and microbial origin with polyethyleneglycol increases the activity of the enzyme used in the treatment of mice bearing inoculated lymphosarcoma 6C3HE and other tumors [88], and it was demonstrated that the survival time (20–24 days) was more than twice that for animals treated with native enzymes. At present, clinical studies with some preparations mentioned above are in progress.

Interesting data are presented in [89, 90]. In these studies, superoxide dismutase was immobilized on polyethyleneglycol. The immobilized superoxide dismutase demonstrated good protective action against lung injuries caused by oxygen metabolites and did not affect the normal bactericidal capacity of neutrophiles.

The modification of porcine pancreatic elastase with activated derivates of polyethyleneglycol-2-O-methoxypolyethyleneglycol-4,6-dichloro-S-triazine yielded immobilized enzymes, demonstrating 35% of the native enzyme activity towards low molecular weight substrates and 17% of the initial activity towards macromolecular substrate (casein [91]). It is important that the modified enzyme completely lost its ability to interact with rabbit anti-elastase antiserum. Besides, it was less inhibited by natural macromolecular inhibitors from the blood, like, α_2-macroglobulin.

Arginase from bovine liver, modified with polyethyleneglycol, also demonstrates increased half-life in the circulation of mice [92]. Enzyme activity after intravenous administration can be discovered in the blood after 72 hours in the case of immobilized enzyme injection, whereas the native enzyme completely clears from the circulation in 5 hours.

Adenosine deaminase and superoxide dismutase increase their stability and half-life in the circulation of experimental animals and decrease their immunogenicity after modification with polyethyleneglycol [93]. Polyethyleneglycol-modified adenosine deaminase can be used for the treatment of adenosine deaminase deficiency [93a]. In two treated children, the modified enzyme was rapidly absorbed after intramuscular injection and had a half-life in plasma of 48 to 72 hours. Weekly doses of approximately 15 U per kilogram of body weight maintained plasma adenosine deaminase activity at two to three times the level of erythrocyte adenosine deaminase activity in normal subjects.

Similar improvement of stability was found with modified L-asparaginase [94], β-glucuronidase [95] and β-glucosidase [96].

Modified forms of proteolytic enzymes, the natural substrates of which are different high molecular weight substances, are of great potential clinical importance. It has been demonstrated that sometimes the binding of proteases with soluble polymers leads to the formation of derivates which possess an increased stability and decreased affinity toward natural enzyme inhibitors or suppressors. Hence, Lindenbaum et al. [97] have shown a correlation between a decrease in acute toxicity of dextran-bound enzymes and a decrease in immobilized enzyme affinity for natural macromolecular inhibitors. The binding of trypsin and terrilytin with serum albumin results in preparations with decreased acute toxicity and low inhibition by blood inhibitors (by 2–4 orders of magnitude less than for native enzymes [98–100]). Among other examples: kallikrein, which is covalently bound with activated polyvinylpyrrolidone, being capable of decreasing canine blood pressure for a considerably longer time than the native enzyme, with a simultaneous sharp decrease in side effects [101]; and lysosomal hydrolase β-D-N-acetylhexoseaminidase (used for the treatment of hereditary Tay-Sachs disease) bound with activated polyvinylpyrrolidone, possessing an increased stability towards exogenous proteases and having a noticeably increased clearance time from the circulation in experimental animals (the native enzyme completely disappears from the circulation in one hour, whereas the immobilized enzyme activity in the blood even after 5 hours is still more than 20% of the initial value [102]).

Proteolytic enzymes of broad specificity—trypsin and α-chymotrypsin, which are used in the treatment of burns and festers, form macromolecular complexes (m.w. 150 000) and acquired increased stability against autolysis [103] after interaction with activated polyvinylpyrrolidone (m.w. 10 000).

Immobilized derivates of terrilythin contain equimolar amounts of the protein and the carrier, maintain high activity towards casein and do not after the main catalytic parameters [104, 105]. The stability of the enzyme in the conjugate is noticeably increased.

A soluble product obtained by immobilization of ocrase (a proteolytic enzyme) on acrylamide-acrylic acid copolymers via the activation of carboxylic groups with water-soluble carbodiimide has been described [106]. Indolyl-3-alkane α-hydroxylase, which possess antitumor activity, was modified with polyacrylic or polymethacrylic acid [107]. As a result, the enzyme activity in

Table 6. Antitumor activity of nuclease-dextran-40 in mice with Erlich sarcoma (adapted from [111])

Preparation	Administration method	Single dose (units · 10^3)	Intervals between injections (hrs)	% of tumor growth inhibition	ALT*, % from control
Nuclease-dextran-40	intraperitoneal	3	12	70–85	0
	intraperitoneal	6	24	50–55	0
Native nuclease	intraperitoneal	3	12	20–25	0
	intramuscular	3	12	+30	15
Nuclease-dextran-40	intramuscular	3	12	20–25	30
	intramuscular	20	12	40–50	40–50
Native nuclease	intramuscular	20	12	+3	—

Note: The duration of the treatment was always 10 days.
* ALT—average life time.

physiological conditions was increased due to the shift of the pH-optimum from pH 3.5 to pH 7.0, despite a decrease of the half-life in the circulation.

Ribonuclease [107] and trypsin [108], immobilized on non-immunogenic poly-DL-alanine, demonstrated decreased antigenicity and immunogenicity. L-asparaginase, modified with the same carrier, prolonged circulation time in mice and rats from 3–5 to 20–26 hours and was less antigenic than the native enzyme.

In experimental cancer chemotherapy in mice, nucleodepolymerases (nucleases) of microbial origin stabilized by covalent binding with soluble aminobenzoyloxymethyldextran (via diazocoupling) possess a 3-4-fold higher activity in the treatment of carcinoma than the native preparations [110, 111], see also Table 6.

Mannanase, collagenase and glucuronidase immobilized on dextran exhibited an increased therapeutic antirheumatic activity after parenteral administration in rats with experimental adjuvant arthritis [112].

The effect of therapeutic enzyme immobilization on soluble polymers on their circulation time in experimental animals and their principal kinetic parameters has been analyzed [93] and the data available are summarized in Tables 7 and 8.

Frequently, the chemical nature of a carrier defines the character of the polymer effect on the immunogenicity of conjugated protein antigens. It has been shown more than once that the use of polysaccharides and polyethyleneglycol [79, 85, 86, 113] leads to a decrease in immunological and allergic reactions towards proteins after administration of immobilized proteins (catalase, streptokinase, bovine serum albumin).

The properties of catalase, modified with polyethyleneglycol PEG-1990. 1900-molecular weight of PEG may be considered as an example (Table 9). Some interesting details on the mechanism of the polyethyleneglycol and monomethoxypolyethylene glycol effect on the antigenicity of proteins have been described [113a, b]. It was shown in these studies that antigens, such as ovalbumin, dog serum albumin, bovine pancreatic ribonuclease and others could be rendered

Table 7. Effect of various polymers on the half-life of circulating enzymes. (From Abuchowski A. and Davis F.F., in: Enzymes as Drugs, Holcenberg, J.S. and Roberts, S., Eds., John Wiley and Sons, New York, 1981, Chap. 13, with permission)

Enzyme	Polymer	Clearance from blood (% remaining/by appropriate time or time of half-life, $t_{1/2}$)
β-D-N-Acetyl hexosaminidase A	Poly(N-vinylipyrrolidone)	40%/1 hr, 21%/5 hr, 14%/48 hr
β-D-N-Acetyl hexosaminidase A	None	0%/1 hr
Carboxypeptidase G	Dextran	$t_{1/2} = 17$ hr
Carboxypeptidase G	None	$t_{1/2} = 3.5$ hr
Arginase	Dextran	$t_{1/2} = 12$ hr
Arginase	None	$t_{1/2} = 1.4$ hr
Arginase	Polyethylene glycol	15%/72 hr
Acinetobacter glutaminase-asparaginase	Glycopeptide	$t_{1/2} = 16$ hr[a], 8.2 hr[b], 19 hr[c]
Acinetobacter glutaminase-asparaginase	None	$t_{1/2} = 1.1$ hr[a], 1 hr[b], 4.3 hr[c]
Pseudomonas 7A glutaminase-asparaginase	Asialoorosomucoid	$t_{1/2} = 7$ min (reached liver)
Lysozym	Desialited fetiun	reached liver in 10 min
Pseudomonas 7A glutaminase-asparaginase	None	$t_{1/2} = 7$ hr
E. coli asparaginase	Lactose (17 residues/mol)	Targeted to liver within 10 min
E. coli asparaginase	None	$t_{1/2} = 4$ hr
Indolyl-3-alkane α-hydroxylase	Polymaleic acid	$t_{1/2} = 2$ hr[a]
Indolyl-3-alkane α-hydroxylase	None	$t_{1/2} = 6$ hr[a]
E. coli asparaginase	Poly-DL-alanine	$t_{1/2} = 21$ hr[a]
Erwinia carotovora asparaginase	Poly-Dl-alanine	$t_{1/2} = 36$ hr[a]
Erwinia carotovora asparaginase	None	$t_{1/2} = 5$ hr
Uricase	Albumin (homologous)	$t_{1/2} = 26$ hr
Uricase	None	$t_{1/2} = 6$ hr
Uricase	Polyethylene glycol	20%/47 hr
Catalase	Polyethylene glycol	10%/48 hr
Catalase	None	6%/6 hr
Adenosine deaminase	Polyethylene glycol	$t_{1/2} = 30$ hr
Adenosine deaminase	None	$t_{1/2} = 0.5$ hr
Superoxide dismutase	Polyethylene glycol	40%/72 hr
Superoxide dismutase	None	6%/4 hr

[a] Mice.
[b] Rats.
[c] Rabbits.

nonantigenic, nonallergenic and tolerant by conjugation with polyethylene glycols or monomethoxypolyethylene glycols of different molecular weights (2000, 5000, 10000 and 20000).

This phenomenon is regarded as due to the decreased ability of modified enzymes to stimulate the formation of antibodies and bind the circulating ones due to steric hindrances of the normal protein-receptor or protein-antibody interaction, created by the presence of a polymeric carrier matrix. After immobilization of proteins (enzymes) on copolymers of D-glutamic acid and D-

Table 8. Effect of various polymers on enzyme parameter (from Abuchowski A. and Davis F.F., Soluble polymer-enzyme adducts, in: Enzymes as Drugs, Holcenberg J.G. and Roberts G., Eds., John Wiley and Sons, New York, 1981, Chap. 13, with permission)

Enzyme	Polymer	Heat stability	Resistance to protease[a]	pH optimum[a]	Km[a]
E. coli asparaginase	Lactose (21 res/mol)	Increased	ND[b]	ND	ND
E. coli asparaginase	Lactose (8 res/mol)	Unchanged	ND	ND	ND
E. coli asparaginase	Polyethylene glycol	ND	ND	ND	Increased
E. coli A-1-3- asparaginase	Polyethylene glycol	ND	Yes	ND	ND
Acinetobacter glutaminase-asparaginase	Glycopeptide from human fibrin	Increased	None	Unchanged	Unchanged
E. coli asparaginase	Poly-DL-alanine	Decreased	Yes	ND	Unchanged
E. carotovora asparaginase	Poly-DL-alanine	Increased	ND	ND	Unchanged
Lysozyme	Dextran	Increased	ND	ND	ND
β-glucosidase	Dextran	Increased	ND	ND	ND
α-amylase	Dextran	Increased	ND	ND	ND
Trypsin	Dextran	ND	Yes	ND	Unchanged
Trypsin	Polyethylene glycol	ND	Yes	ND	ND
Trypsin	Polyvinylpyrrolidone	ND	Yes	ND	Increased
Catalase	Dextran	Increased	Yes	ND	ND
Catalase	Polyethylene glycol	Unchanged	Yes	Unchanged	ND
β-D-N-acetylhexosaminidase A	Polyvinylpyrrolidone	Unchanged	Yes	Unchanged	Unchanged
Indolyl-3-alkane α-hydroxylase	Polymaleic acid	ND	ND	Alkaline shift	Slight increase
Uricase (C. utilis)	Polyethylene glycol	ND	Yes	Alkaline shift	Slight increase
Uricase (hog liver)	Polyethylene glycol	ND	Yes	Alkaline	Increased
Uricase	Albumin	Increased	Yes	ND	ND
Phenylalanine ammonialyase	Polyethylene glycol	ND	Yes	Unchanged	Increased
Arginase	Polyethylene glycol	ND	Yes	ND	Increased

[a] Relative to the native enzyme.
[b] Not determined.

Table 9. Immunodiffusion studies using catalase, and catalase-modified to varying degrees with PEG-1900, against rabbit antisera to each antigen. Antisera were prepared in rabbit pairs by the intravenous injection of catalase, or catalase to which increasing amounts of PEG-1900 were attached. Center wells contained 10 ul of antiserum and peripheral wells contained 10 ul of 1 mg/ml of solution of catalase or modified catalase in 0.01 M phosphate-buffered saline, pH 7.3. Results are expressed as estimated relative amount of precipitation for each antiserum-antigen pair. Tests were carried out in quadruplicate. + + + indicates greatest precipitation; – indicates no observable precipitation. (From Abuchowski A., McCoy J.R., Palczuk N.C., van Es Th., and Davis F.F., J. Biol. Chem. 202, 11, 3582, 1977, with permission)

	Antigen				
Antiserum to	Catalase	PEG-catalase (13%)	PEG-catalase (19%)	PEG-catalase (37%)	PEG-catalase (43%)
Catalase	+++/+++	++/++	++/++	+/++*	+/+
PEG-catalase (13%)	++/++	++/++	++/+	–/–	–/–
PEG-catalase (19%)	++/+	++/–	++/–	++/–	–/–
PEG-catalase (37%)	+/+	–/++	–/++	–/++	–/++*
PEG-catalase (43%)	–/–	–/–	–/–	–/–	–/–*

* Results checked by complement fixation.

lysine conjugates are formed which possess the ability to suppress specific immune responses, leading to the prevention of immunoglobulin E formation responsible for local and systemic allergic reactions [114].

However, conversely, it has been found that some synthetic polyelectrolyte carriers are capable of actively increasing immunogenicity provoking an increased formation of antibodies in response to the administration of proteins bound with such carriers in an electrostatic complex which is capable of being destroyed under physiological conditions [115]. Highly antigenic conjugates obtained by binding weakly antigenic proteins to stimulating carriers can be used for the preparation of synthetic vaccines of a new type.

Generally speaking, immunological properties of immobilized therapeutic enzymes have been unsufficiently investigated. Studies on immunological properties of polymeric carriers (dextran) and their conjugates with enzymes are still rather rare [116].

Modification by water-soluble polymers has been used to prepare different therapeutic proteins. Efforts to obtain hemoglobin polymeric derivatives (semi-synthetic blood substitutes) are of great importance. It has been shown that hemoglobin binding with bromo, amino or aldehyde derivates of dextrans gives products which possess 3 to 20 fold higher circulation half-lifes compared to the native hemoglobin, with its functional properties being unchanged [117–119]. It is encouraging to observe that experiments carried out an animals with hematocryte values of less than 2% showed that the transfusion of dextran-hemoglobin complexes allowed animals to survive and to recover fully without additional enrichment of the air with oxygen. Analogous data were obtained with hemoglobin immobilized on dialdehyde dextrans with a molecular weight of 45 000 and 25 000 [120].

A polymer of hemoglobin, obtained by cross-linking with glutaric dialdehyde, has also been described, demonstrating functional activity and increased stability [121].

The problem with the application of protein inhibitors of enzymes is very closely connected with clinical application of the latter. One of the inhibitory polypeptides which is widely used in clinical practice is a polyvalent inhibitor of proteolytic enzymes from bovine pancreas. The pancreatic inhibitor effectively inhibits kallikrein, trypsin, chymotrypsin, plasmin, plasminogen activators, blood clotting factors, tissue and leucocytic proteases. It is used in the treatment of allergies, shock, sepsis, acute pancreatitis, mechanical and thermal injuries, artrosoarthritis, primary hyperfibrinolytic bleeding and in the treatment of myocardial infarction. Unfortunately, the half-life of the pancreatic inhibitor in the circulation (depending on the animal species) is only between 10 and 70 min, which requires increased therapeutic doses. To prolong its action, a number of derivates modified by soluble polymers have been developed. Inhibitor preparations, which possess a prolonged half-life (in rats, a 10-fold increase), and unchanged activity have been described [122, 123]. The application of a soluble high-molecular-weight pancreatic inhibitor in dogs with experimental acute pancreatitis increases animal survival 1.5 to 2 times. Introduction of sugar moieties which have receptors in liver cells into the carrier permits direct accumulation of the modified inhibitor in the liver [124].

Apart from the pancreatic inhibitor of kininogenases, some other inhibitors of the kallikrein system have been immobilized on water-soluble carriers [125], including the nonapeptide bradykinin and a synthetic hexapeptide inhibitor of angiotensin-I-converting enzyme—SQ 21541. It has been shown in vitro that the latter peptide, modified with soluble dextran, expresses 24% of inhibitory activity of the native peptide [125]. Bradykinin bound with dextran activated by cyanogen bromide or periodate maintains its immunological activity, and retains about 6% to 25% of its biological activity. Possibilities of clinical application of other immobilized protease inhibitors have been studied [112].

In a number of cases, the comparison of the biological activity of native enzymes and their macromolecular derivates allows elucidation of some of the important aspects of the functioning of physiologically active substances (low molecular weight drugs, hormones and neurotransmitters), a detailed review in Refs. [126, 127].

It is expedient to include into this section the data on enzymes stabilized without the use of a carrier, i.e., by chemical modification of proteins or by the introduction of intramolecular cross-links into a protein which under the unfolding of an enzyme molecule (see review [Ref. 128]). The principle of stabilization of single-chain and subunit enzymes by intramolecular cross-linking with the help of bifunctional reagents [129, 130] is illustrated in Fig. 4. Dialdehydes, diacids, diimidoesters etc. can be used as bifunctional cross-linking agents. The preparations obtained possess higher stability against unfolding and the action of proteases. The clearance rate and inhibition activity of these modified enzymes should be the same as for native enzymes. Nevertheless, the approach can be applied to enzymes which need to penetrate a cell or bind with receptors in the cellular membrane, where the presence of a polymeric carrier may hinder these processes.

A relatively new method of stabilizing therapeutic enzymes is their immobilization on substances which are either characteristic of the organism itself (e.g. proteins), or even possess their own useful biological activity and improve the properties of an immobilized enzyme. Examples include the immobilization of lysozyme and asparaginase on collagen, the antigenic properties of which may be decreased by preliminary limited proteolysis [132, 133]. Enzyme immobilization on fibrin, preliminarily treated by heating to reduce antigenicity, has been described [134].

The immobilization of α-galactosidase, uricase and L-asparaginase on albumin ensures considerable protection of the enzymes against thermal and proteolytic inactivation in vitro [135, 136]. Trypsin modified with albumin via glutaric dialdehyde is less sensitive to autolysis and to the action of serum inhibitors than the native enzyme [137]. Poznansky [138, 139] has shown in animal experiments that the circulation half-life of uricase/albumin conjugates in dogs after intravenous injection totals 26 h as compared with 4–6 h for native uricase. In the same paper it was shown that with the native enzyme, the first injection decreased the amount of uric acid in the plasma by 35% with a subsequent return to the initial level with the half-time of approximately 24 h; the second injection was less effective; and the third caused an anaphylactic shock. This fact points to the fast formation of antibodies against uricase. In contrast, repeated injections of albumin-immobilized uricase remain as effective as the first, with no immunologic complications.

One more example of using "natural" biological carriers for the immobilization of therapeutic enzymes may be the glycosylation of glutaminase-asparaginase with glycopeptides from human fibrin and γ-globulin [140]. The enzyme in such a conjugate is not only more stable than the native asparaginase, but its removal from the circulation of experimental mice, rats and rabbits is 15 times slower.

Finally, soluble stabilized enzymes can be obtained by intermolecular cross-linking with bifunctional reagents. Thus, cross-linking of β-galactosidase (the absence of which causes Fabry's disease) leads to noticeable enzyme stabilization [141]. The trimer of pancreatic ribonuclease has a 12-times longer half-life in the circulation of rats than that found with the native enzyme [142].

One should bear in mind that the formation of protein conjugates, having a great number of groups capable of simultaneous interaction with cell surfaces, may lead to an increase in their capture by cells. Avid capture of ribonuclease dimers and polymers have been demonstrated in both hepatoma cell cultures [143] and nephrectomized rats [144]. Increased capture of antibody aggregates in kidney capillaries has also been demonstrated [145]. Nevertheless, according to [146], soluble cross-linked enzyme polymers may have a great future in enzyme therapy.

The studies mentioned above show that the field of soluble, stabilized therapeutic enzymes is wide spread and includes practically all groups of enzymes used as drugs. The principles governing changes of enzyme properties after immobilization have been established and the most convenient carriers for use in medicine have been selected, such as polysaccharides and polyethylene glycols. At the same time, one can easily notice that the majority of studies have been in the laboratory and only few of the preparations investigated are close to become real drugs.

4.1.1.2 Thrombolytic Therapy with Immobilized Enzymes

A special position is occupied by immobilized thrombolytic enzymes. This particular field includes a great number of studies; several examples for the clinical use of immobilized thrombolytic enzymes already exist.

At present, thrombolytic therapy seems to be extremely important in the treatment of cardio-vascular and certain other diseases. It has been shown to be very effective in the treatment of myocardial infarction, pulmonary embolism, arterial occlusion and deep vein thrombosis [147–149].

Thrombolytic enzyme therapy has two principal approaches: a) the use of the enzyme (or its zymogen) which is responsible for fibrin degradation in physiological conditions, and b) the use of proteolytic enzymes capable of specific transformation of the appropriate zymogen into fibrin-degrading enzymes.

Several factors limit the everyday clinical use of thrombolytic enzymes. Among them are the high cost and relatively low availability of pure enzymes, which is especially true in the case of urokinase and tissue plasminogen activator. On the other hand, streptokinase, which is at present the most widely used therapeutic thrombolytic enzyme, has been isolated from cultural media of β-hemolytic streptococci and demonstrates a relatively high antigenicity.

The enzymes are also sensitive to the action of endogenous degrading enzymes and various natural inhibitors; they do not express specific affinity towards the affected area. To achieve high local therapeutical concentrations, one needs to increase the total quantity of the administered enzyme. Thrombolytic enzymes also demonstrate a pronounced depressive action on the coagulative component of hemostasis.

In a clinical situation, this results in the necessity of prolonged and continuous intravenous administration, frequent allergic reactions and hemorrhagic and rethrombotic complications. In 80% of cases, hemorrhagic complications have been registered, in 30–60%—reccurent thromboses and embolism due to thrombi fragmentation, 30% of cases are accompanied with allergic reactions [150, 151]. There are also difficulties in controling and carrying out combined therapy with heparin.

At present, the following thrombolytic enzymes are used in clinical practice:

Plasminogen and plasmin [152]. Human plasminogen is a single chain glycoprotein (a carbohydrate content of approx. 2%) with a m.w. of about 90 000. In the native form, plasminogen has a N-terminal glutamic acid (so-called "Glu-plasminogen") and can be easily converted into "Lys-plasminogen", containing N-terminal lysin, valin or methionin. Both zymogens are proteolytically converted into the active form—plasmin, which is a trypsin-like enzyme, consisting of two chains bound together with two S–S bridges and containing His-602, Asp-645 and Ser-740 at the active site. Plasmin from human blood has limited use in medicine and the therapeutic enzyme is usually obtained by trypsin activation of purified porcine plasminogen.

Urokinase [153, 154]. This serine protease can be isolated from human urine or kidney tissue cultures and exists in two molecular forms (m.w. approx. 33 000 and 53 000). Urokinase is very selective in its action on proteins, the conversion of plasminogen into plasmin being the only known reaction catalyzed by this enzyme.

Tissue plasminogen activator (TPA) [155, 156]. TPA is also a serine protease with a m.w. of approx. 68 000, which is synthesized in endothelial cells and present in most or all human tissues. TPA has some homology with urokinase and plasminogen and demonstrates high affinity for fibrin. The physiological role of TPA is the activation of plasminogen into plasmin in the region of thrombus formation, where the activity of TPA is greatly enhanced by the presence of fibrin.

Streptokinase [157]. Streptokinase is the neutral proteinase from *Staphylococcus aureus* which catalyzes the conversion of plasminogen (the only known substrate for streptokinase) into plasmin. Streptokinase itself does not catalyze the proteolytic cleavage of plasminogen, but forms preliminary complexes with circulating plasmin or plasminogen, which act as catalysts for plasmin formation. The m.w. of streptokinase is 45 000–47 000. The preformed native or acylated complex of streptokinase with plasminogen can also be successfully used as a thrombolytic agent [158].

Attempts were made to enhance the efficacy of thrombolytic enzymes (the rate of thrombus lysis), simultaneously decreasing the total dose of enzyme needed and its side effects. In this respect, enzyme administration via a catheter directly into a vessel, one of the branches of which is occluded with the thrombus, is of particular interest. Intracoronary injection of urokinase in the treatment of patients with acute myocardial infarction demonstrated enhancement of therapeutic efficacy [159].

The development of new thrombolytics has aroused interest for the use of thrombolytic enzymes in the immobilized state [160].

In the first studies in this field streptokinase was chemically bound to carboxymethyldextran or dextrans modified with cyanurchloride, cyanogenbromide and various carbodiimides [161, 162]. These preparations of immobilized streptokinase contained the enzyme and the carrier in a weight ratio of 1:10–200 and preserved the ability to lyse thrombi, exhibiting greater thermostability than the native streptokinase. Soluble urokinase derivates were obtained by enzyme interaction with BrCN- or periodate-activated dextran with a m.w. of 20 000. 100 to 10 000 IU of the enzyme were bound per gram carrier and maintained its activity, demonstrated increased stability and prolonged half-life in the circulation. Enzyme to polymer binding was achieved directly or via a spacer group containing 1 to 3 carbon atoms [163, 164].

Streptokinase, plasmin and urokinase immobilized on dextran activated with partial oxidation by periodates have been described [164–167]. This activation method allows to avoid the use of highly toxic activators such as cyanurchloride or cyanogenbromide. In all cases immobilized enzymes separated from non-bound protein by chromatography on Sepharose 4B demonstrate unchanged specific activity and enhanced stability. For example, in case of plasmin immobilized on aldehydedextran [164] a preparation was obtained containing ca. 100 mg of bound protein per 1 g carrier; preservation of enzyme specific activity towards the low molecular weight substrate—N-benzoyl-L-arginine ethyl ester, was 85%. The specific fibrinolytic activity of the preparation was 22 U/mg.

Other soluble polymers are also suitable carriers for immobilization of thrombolytic enzymes. Thus, urokinase was immobilized on a soluble copolymer of acrylamide with acrylic acid [168]. The content of reactive residues of acrylic acid in the polymer may vary from 1 to 15%. Urokinase binding to the polymer with acrylic acid content of about 10% mol and a m.w. of approx. 70 000 was achieved with 1-ethyl-3(3-dimethylaminopropyl) carbodiimide as a coupling agent. The preliminary optimization of reaction conditions of protein binding on carboxyl-containing polymers via carbodiimide permits a binding of up to 180 mg of the enzyme per g carrier [169]. The immobilization yield was 90%; activity preservation (using N-acetyl-L-glycyl-L-lysine methyl ester as the specific substrate) was 95%. The process was performed in two steps: the carrier activation with carbodiimide in the buffer-free medium at pH 4.0 (10 min incubation) was followed by protein immobilization on the activated carrier (1 h incubation at 20 °C, in borate buffer pH 8.5).

The enhancement of urokinase stability may be also achieved by a rather simple way—the formation of electrostatic urokinase complexes with Na dextransulphate [170, 171] or heparin [172], although a long-term in vivo existence of such complexes is unlikely, as they can be destroyed by mere increase of the ionic strength of the solution. Prolongation of urokinase lifetime in the blood flow without essential decrease in fibrinolytic activity may be achieved by the removal or partial degradation of the carbohydrate part of the urokinase molecule [173].

Streptokinase immobilization on copolymer of N-vinylpyrrolidone and acrolein diethylacetal was used for the comparative study of conformational properties and conformational stability of native and immobilized enzymes [174]. It was shown that the conformational stability of the immobilized enzyme in the presence of urea sufficiently increases as compared with the native enzyme, the thermoinactivation rate of the modified streptokinase also being lower than that of the native enzyme.

Urokinase immobilization on polyalkylenglycoles with a m.w. of 200 to 20 000 [175] and streptokinase immobilization on polyethylenglycol with a m.w. of 2 000, 4 000 and 5 000 have been described [176] in a detailed study of kinetics of streptokinase modified with polyethylenglycol during plasminogen activation. In case of polyethyleneglycoles activated with carbonyldiimidazole, the products obtained exhibit practically unaltered reaction activity in plasminogen activation. It is interesting to note that the molecular weight of polyethyleneglycole has a distinct impact on adduct properties: corresponding derivates of streptokinase exhibit somewhat different catalytic properties, different stability to plasmin action etc. This result points to the necessity of optimization the properties of immobilized enzymes not only according to the type of carrier but also according to its molecular weight. Modified enzymes, irrespective of molecular weight of carriers, exhibit a sharply decreased ability for interaction with antibodies against native streptokinase. The authors have also studied the clearance of radiolabeled streptokinase/polyethylenglycol in mice and demonstrated that the modified enzyme preserves its activity in the blood flow longer when compared with the native enzyme; inactivation and clearance processes proceed in parallel (see Figs. 19–21).

From the practical point of view, the most important results were obtained with streptokinase immobilized on soluble activated dextran with a m.w. of 35 000–50 000 [166, 167]. With streptokinase all the drawbacks of thrombolytic enzyme therapy are maximally expressed: the enzyme exhibits low stability and short half-life in the circulation; it requires continuous and prolonged intravenous administration; it deeply affects the coagulation system

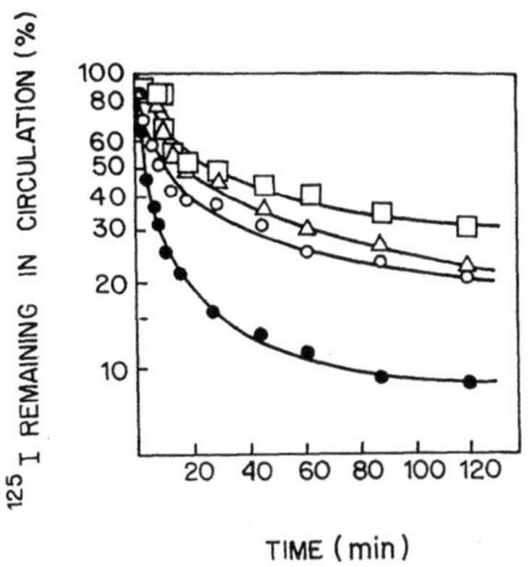

Fig. 19. Plasma elimination of PEG-modified ^{125}I-streptokinase. The percentage of the injected radiolabel remaining in the circulation was calculated relative to an initial time point drawn 5 s after injection. Streptokinase was unmodified (●), or modified with PEG-2 (□), PEG-4 (△), or PEG-5 (○) (Reproduced with permission from Rajagopalan S et al. (1985) J Clin Invest 75:413)

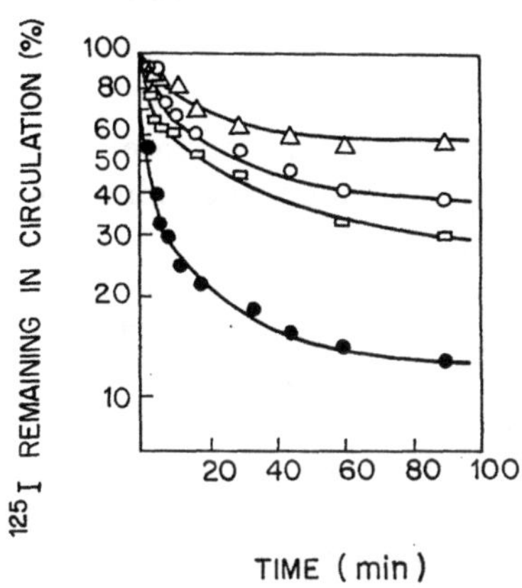

Fig. 20. Plasma elimination of streptokinase modified with PEG-2 under alternative reaction conditions. Clearance of native streptokinase (●) is compared with the clearance of streptokinase by protocol A (□), protocol B (△), and protocol C (○). PEG-streptokinase coupling. Activated PEG was reacted with streptokinase (1 uM) in 10 mM sodium borate buffer, pH 8.5, at 4 °C for 72 h with 40 mM of activated PEG (protocol A). Activated forms of PEG spontaneously hydrolyze in water and this may affect the efficiency of protein coupling. To determine whether more extensive derivatization would cause loss of protein activity protocol A for reacting streptokinase was modified by incubation off streptokinase with 80 mM of activated PEG under the above conditions (protocol B). In one experiment, the streptokinase reacted with 40 mM of PEG for 48 h and then with 80 mM of PEG for an additional 48 h (protocol C). As a control, streptokinase was also incubated with nonactivated PEG under identical conditions (Reproduced with permission from Rajagopalan S et al. (1985) J Clin Invest 75:413)

causing numerous hemorrhagic or rethrombotic complications; it is highly allergic because of the usually high titers of antistreptococcal antibodies in the human blood. Aldehyde-dextran modified streptokinase seems to be a promising alternative to the native enzyme. The immobilized enzyme is available in the USSR under the trademark "STREPTODEKAZA"^{R} and is used in the treatment of acute myocardial infarction, acute pulmonary artery thromboembolism, peripheral arterial and deep vein thrombosis and hemophtalm [179–181].

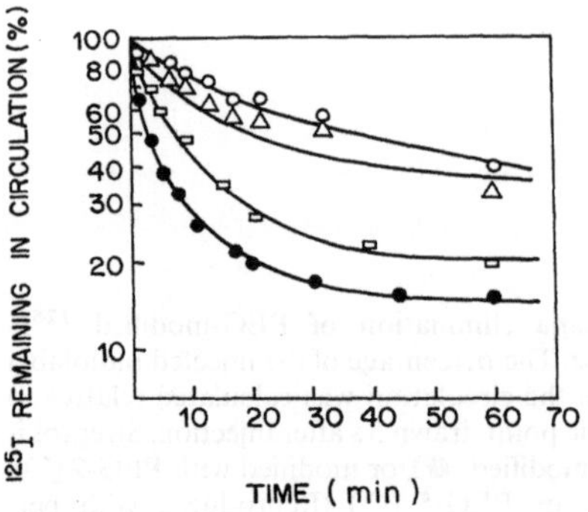

Fig. 21. Plasma elimination of an activator complex formed with ^{125}I-plasminogen and streptokinase modified with PEG-2 under alternative reaction conditions. The activator complex contained ^{125}I-plasmin and unmodified streptrokinase (●) or streptokinase modified by protocol A (□), protocol B (△), or protocol C (○) (see legend to Fig. 20) (Reproduced with permission from Rajagopalan S et al. (1985) J Clin Invest 75:413)

Streptodekaza has a prolonged half-life in the circulation (an increase in the fibrinolytic activity in the blood can be observed even 80 hours after administration) and practically causes no complications.

The treatment of thromboses and emboli with Streptodekaza, the analysis of results and prospects of application. Analyzing results of the treatment with Streptodekaza in 150 patients (54 patients with myocardial infarction, 20 with thromboembolism of pulmonary artery branches; 12 with chronic venous thrombosis, 54 with intravitreal hemorrhage, 10 with retinal vein thrombosis), the following indices were taken into consideration: clinical efficacy, tolerance, complications, specific effect on hemostasis. To evaluate the advantages of Streptodekaza its effect was compared with the effect of the native enzyme.

The mode of the treatment. In absence of contraindications for the use of thrombolytic therapy, 300 000 IU of the drug were diluted with 2 ml physiological solution and injected intravenously. This simultaneously presented a dose neutralizing probable antibodies and a biological test. In one hour (when no side effects were observed) a single injection of 2 700 000 IU of Streptodekaza was given intravenously during 3–5 minutes in 20 ml physiological solution or Reopolyglukin (the USSR trade mark for a solution of dextran 20 000 in physiological solution). The total therapeutic does of the enzyme was 3 000 000 IU, in some cases 6 000 000 IU. After systemic treatment with Streptodekaza, heparin and antiaggregating drugs were prescribed, beginning with the first day of the therapy.

In opthamological practice, Streptodekaza was injected subconjunctively, retrobulbarly, and rarely intravitreally. The usual dose was 30 000–45 000 IU (0.2–0.3 ml solution), the number of injections per treatment course varied from 1 to 10 (depending on the kind of pathology and thrombolysis efficacy) with an interval of 3–5 days. None of the native thrombolytic enzymes exhibited similar advantages for treatment in the indicated doses [181–183].

Streptodekaza efficacy in the treatment of acute myocardial infarction. Significant differences were discovered in the course of disease with Streptodekaza treatment when compared with the efficacy of the treatment with heparin only or with heparin and streptokinase (see Table 10).

The frequency of specific complications after treatment with Steptodekaza was much lower than for streptokinase: allergies and hemorrhagies practically disappeared, and thromboemboli were considerably less frequent. The simplicity of therapy, combined with its safety, are very important for the wide application of thrombolytic therapy in practice.

The treatment with Steptodekaza of massive thromboembolism of the pulmonary artery. Improvement of clinical state (diminution of dyspnea, cyanosis, tacchycardia, ECG signs of the overloading of the right heart) were observed in 65% of the patients even after one or two days after immobilized enzyme administration. Thrombolysis was especially active in the first 2 weeks. It was proved by data on the state of the pulmonary artery system obtained by angiopulmonography, measurement of pressure in the right heart, and lung scans. Perfusion in the lungs increased up to 35% from initial values at the end of the second week.

According to angiopulmonography data, recovery of completely occluded vessels was observed in 33% of patients; however, partial thrombolysis was always observed. A considerable improvement of blood circulation in the pulmonary artery (almost its normalization) was observed already on the 6-th day after a single injection of 6 million U of the drug.

The maximal effect was achieved in patients with fresh thromboembolism and non-occlusive type of lesions; this is typical for treatment with any kind of thrombolytic drug [184–186].

The efficacy of Streptodekaza application in ophthalmology. Streptodekaza was injected intrabulbarly in hemophtalm treatment. A clinical effect was achieved on the 1–2 day, lysis of the blood clot was observed on day 7–8.

Table 10. The comparison of the clinical efficacy of Streptodekaza/Heparin (1), Streptokinase/Heparin (2) and Heparin (3) administration in the treatment of patients with acute myocardial infarction

	Patient groups		
	1 n = 37	2 n = 30	3 n = 19
Clinical criteria	Frequency, %		
Relief of pain syndrome	83.8	63.9	50.0
Fast ECG dynamics	67.6	45.8	16.0
Prolonged course and secondary infarction	5.4	13.3	15.8
Circulatory insufficiency (%)			
a—on admission	54.9	42.8	37.8
b—on the 14th day of disease	3.6	22.8	20.0

Disappearance of hiphema (hemorrhage in the anterior eye chamber) was observed in 100% of the cases. Vision acuity which was on the level of light sensation only increased up to 0.2–0.7, and sometimes up to 1.0. No complications were observed after single or double Streptodekaza injections. In order to achieve a maximal effect, 3–4 injections were performed sometimes. In the latter case, local allergy was observed (hyperemia, edema of eyelids) which was easily abolished by application of dexametazone ointment.

Streptodekaza demonstrated its efficacy in the treatment of a 4 month old chronic thrombosis of the central retinal vein in 90% of the patients, the acuity of vision increased (in average) from 0.22 to 0.48. Local allergic reactions were observed in 90% of the patients after 2–3 injections. Contrast fluorescent angiography of eyeground vessels after Streptodekaza injections revealed the acceleration of venous flow, perfusion recovery, formation of collaterals, i.e. direct and indirect indications of thrombolysis.

When evaluating the results of the clinical application of Streptodekaza in ophthalmology, one should pay special attention to the simplicity of therapy and its efficacy as compared with existing drugs and surgical treatment of traumatic hemophtalm and thrombosis of the retina [181, 182].

All positive properties of the new drug are the result of chemical modification of the streptokinase molecule with a polysaccharide matrix. The immobilized enzyme represents a complex of the activator-carrier type. The polysaccharide matrix plays a protective-transport function, protecting blood fibrinogen from the degradation by plasmin and the active centres of the immobilized enzyme from the effect of natural inhibitors.

To increase the therapeutic activity of thrombolytic enzymes different substances can be used for their modification, stabilization or immobilization which themselves possess useful physiological activity or can somehow improve or facilitate enzyme therapeutic action.

Thus, it was suggested to use heparin as a carrier for proteolytic (thrombolytic) enzymes immobilization, since heparin has low antigenicity, high chain rigidity (which is important for its stabilizing effect on the immobilized protein) and high content of reactive carboxylic groups which can interact with protein amino groups after preliminary activation with water-soluble carbodiimde [187, 188]. On the other hand, it should be also kept in mind that thrombolytic enzyme therapy is usually performed together with heparinization. To obtain preparations exhibiting both thrombolytic and anticoagulant activity together with increased stability under physiological conditions, plasmin and urokinase were immobilized on heparin. For this purpose, the enzyme was incubated at pH 8.2 with heparin after preliminary activation of heparin carboxylic groups with carbodiimide at pH 4.5. Preparations have been obtained containing up to 250 mg of protein per g of heparin (the degree of binding achieved was as high as 95%), with 80 to 85% of the enzyme activity preserved. Immobilization on heparin does not affect the catalytic properties of an enzyme, but greatly increases enzyme stability—the half-time at 37 °C in phosphate buffer, pH 7.4 for native plasmin is approx. 1 hour, but reaches 18 hours for the heparin-modified enzyme.

The anticoagulative activity of heparin conjugated with the thrombolytic enzyme remains unchanged (as measured by the standard test with ADP-induced platelet aggregation).

Another example of thrombolytic enzyme modification with a physiologically active compound, is sodium nitroprusside [188] which is known to be very active antihypertensive agent, the extremely short half-life in the circulation being its main drawback. Non-covalent binding of sodium nitroprusside with urokinase gives preparations exhibiting simultaneously thrombolytic and vasodilatory activities.

After modification of urokinase with sodium nitroprusside, the enzyme retains 85% of its initial exterase activity.

Intravenous injection of free and urokinase-bound sodium nitroprusside in Wistar rats showed that the application of free sodium nitroprusside causes blood pressure to decrease by 40–50% as compared with the initial value. The maximal decrease occurred at 11–24 sec after the injection and the return to the initial level is achieved in about 1 min. In the case of urokinase-bound sodium nitroprusside, these parameters amount to 65–70%, 15–21 sec and 4–7 min, respectively.

Recently, the development of new drugs combining properties of different thrombolytics was initiated. The production of such drugs is based on the ability of various thrombolytics and their fragments to form complexes with each other, which may be considered as the immobilization of one catalytically active protein on another and it is a logical continuation of the use of physiologically active carriers for the immobilization of therapeutic enzymes. Thus, a complex of streptokinase and light B-chain of human plasmin has been described [158]. This complex exhibits higher activity and longer half-life in the blood flow as compared with native streptokinase. The half-life of the complex in the blood flow of experimental dogs was 4 hours, that is at least 2 times longer than for native streptokinase. Streptokinase-plasmin complexes may be additionally modified by soluble polymers, i.e. polyethylenglycol, to impart enhanced stability in the blood flow [190].

A new approach is the construction of a hybrid thrombolytic preparation. A covalent hybrid plasminogen activator was derived from human plasmin fibrin-binding and tissue plasminogen activator catalytic domains [190a]. The hybrid molecules contained 1 mol of each chain, have a m.w. of approx. 92000 and demonstrate about 40% of the specific activity of the parent tissue plasminogen activator (~200000 IU/mg of protein). The hybrid demonstrates also good binding to fibrin and high fibrinolytic activity.

4.1.2 *Artificial Cells*

The so-called "artificial cells" are another form of immobilized therapeutic enzymes, intended for prolonged circulation in the organism. This term, introduced by Chang [191, 192], initially included impermeable and semipermeable microcapsules, into the inner space of which an enzyme (or, in a more general form, any other drug) was encapsulated. More recently, the term included cell ghosts and phospholipid vesicles—liposomes.

Microcapsulated enzymes possess a number of advantages for therapeutic purposes: (1) the microcapsule membrane, on one hand, protects the enzyme from contact with an aggressive physiological medium, and on the other hand, protects normal tissues and organs from undesirable effects of the immobilized drug; (2) considerable amounts of the enzyme may be encapsulated into microcapsules, which increases the enzyme/carrier ratio up to 1000000:1 (see Fig. 22); (3) the

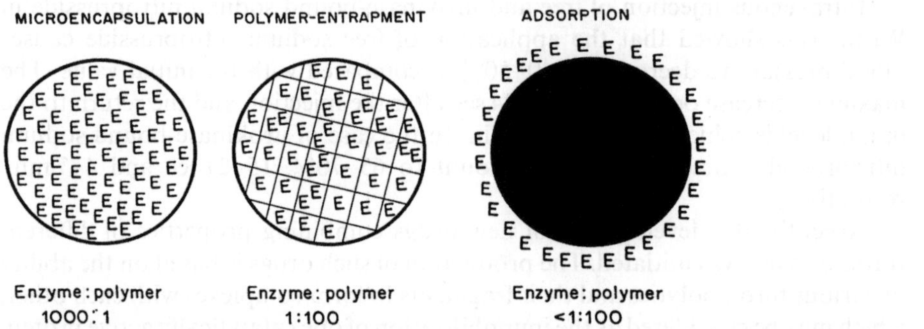

Fig. 22. Relative amounts of matrix support (–) and enzyme (E) for different immobilization methods (Reproduced with permission from Chang TMS (1976). In: Hixson HF Jr, Goldberg EP (eds) Polymer grafts in biochemistry, Marcel Dekker, New York, p 253)

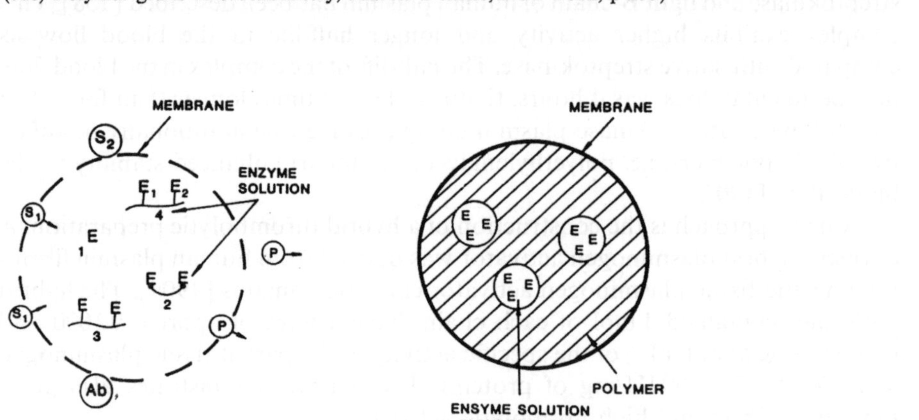

Fig. 23. Schematic representation of enzyme (E) entrapment into microcapsules. **A**—semipermeable microcapsule. Enzyme can be immobilized in the native form (*1*), as intermolecularly cross-linked derivative (*2*), being immobilized on soluble carrier (*3*) or as a component of immobilized multienzyme complex (*4*). Low-molecular-weight substrates (S_1) and products (*P*) can easily penetrate the membrane; at the same time high-molecular-weight substrates (S_2) and antibodies (*Ab*) can not enter the microcapsule. **B**—"double" entrapment of enzyme into polymeric microcapsules (Reproduced with permission from Torchilin VP (1987) Adv Drug Deliv Rev 1:41)

surface area/volume ratio of microcapsules is maximal at a given volume; as a result, there is a fast penetration of substrates and products of the enzymatic reaction via the membrane; (4) an enzyme may be encapsulated into microcapsules and additionally stabilized (intermolecularly or intramolecularly cross-linked enzymes or enzymes immobilized on a soluble polymer, see Fig. 23); and (5) artificial cells can include not only individual enzymes, but also multienzyme complexes, enzyme mixtures or even intact cells. Artificial cells derived from microcapsules, liposomes or cell ghosts have much in common. At the same time, they are usually designed for different purposes and have some principal differences.

Historically, polymeric microcapsules were the first among artificial cells.

4.1.2.1 Properties and Application of Enzymes Immobilized in Semipermeable Microcapsules

The entrapment of a solid, liquid or gaseous substances into an inert, polymeric, hollow spheric membrane is called microencapsulation. If the membrane performs only a protective function and is not permeable for an entrapped substance, the latter can express its activity only upon the destruction of the membrane. In the early 70-s, T.M.S. Chang developed a new type of microcapsules for the immobilization of enzymes [191–198]. These microcapsules had thin membrane, impermeable for encapsulated biomacromolecules, but permeable for low molecular weight substrates and products of the enzymatic reaction. Microrenocapsulated enzymes can not be inactivated by their specific antibodies and macromolecular inhibitors (see Fig. 23). Since that time, many microencapsulation methods have been developed [194–196]. Traditionally, all these methods can be divided into three principal groups (which often overlap): physical methods, methods of phase separation and polymerization (polycondensation) methods.

Physical methods (pulverization by drying, extrusion, centrifugation, encapsulation in a fluidized bed and vacuum coating) are rarely used for enzyme immobilization, because they require the use of extreme conditions which can inactivate the enzymes.

Relatively simple phase separation methods [197–199] are based on the emulsification of an encapsulated substance in the solution of film-forming polymer with subsequent phase separation. As a result of phase separation, the phase of concentrated polymer solution (a "shell") is formed along the interface around particles of the encapsulated substance. The phase separation can be performed in both water and organic solvents. According to recommendations [197], the microencapsulated enzyme can be prepared as follows:

1. Dissolve the enzyme in a buffer solution usually containing another protein, such as albumin, for protection.
2. Separately prepare an organic phase containing a small amount of water-oil emulsifying agent, such as Span 85. The organic phase should be immiscible with water, and commonly used systems are ether, cyclohexane and toluene.

3. Add the aqueous enzyme solution to the organic phase and stir for a predetermined time and speed. The speed of the stirrer determines the size of the microcapsules.
4. To this two-phase mixture, add a second organic solution. This solution contains the polymer and an organic solvent for the polymer in the organic phase originally selected to contain the Span 85. When the second organic liquid is added, small colloids are precipitated out by dilution of the organic solvent.
5. As the solvent evaporates from the organic phase, further precipitation will result in a thicker membrane being formed. The thickness of the membrane is controlled by the amount of polymer added to the organic phase and by the time allowed for phase precipitation.
6. Separate the small microcapsules from the organic phase either by sedimentation or centrifugation and wash with buffered aqueous solution containing a detergent such as Tween 20.

For stabilization during encapsulation, an enzyme can be mixed with stabilizing polymers (bovine serum albumin, hemoglobin, polyethyleneimine). Film-forming polymers and co-polymers of the vinylic series which are readily soluble in non-polar organic solvents are usually used as materials for capsule preparation.

The microencapsulation by polymerization is based on the chemical reaction of polymeric membrane formation on the interface. Unlike in previous methods, in this particular case, semipermeable polymeric membranes are formed directly during the encapsulation process.

When the polycondensation reaction is used for microencapsulation, the monomers to be condensated are solubilized in two immiscible liquid phases. The polycondensation reaction proceeds on the interface and leads to the formation of semipermeable polymeric membranes (see, e.g., 200–202). Usually, the water solution of an encapsulated substance and water soluble monomer are emulsificated in an organic solvent in the presence of a detergent. After that, the solution of the second co-monomer in an organic solvent is added to the emulsion, and the fast polycondensation reaction proceeds on the interface.

Microcapsules, obtained by polycondensation have ultrathin semipermeable membranes. The most widely used are membranes made of polyamides and polyesters [191, 201, 203, 204]. Despite the simplicity of polycondensation methods, some enzymes can not be encapsulated by this way because of their inactivation in the presence of highly reactive monomers or at high pH values.

Routine polymerization and polycondensation usually yield microcapsules from several to several hundreds of microns in diameter. Capsules of the size 20 to 250 nm (so called nanocapsules) [205, 206] are usually obtained by micellar polymerization. For this purpose, a water solution of a substance to be encapsulated is solubilized in a lipophylic solvent in the presence of high concentrations of a surface-active compound. Than, amphiphylic flim-forming monomer is added to the outer phase of the micellar system obtained. It concentrates on the interface and can be polymerized by conventional physical or chemical methods. The capsules are separated from the outer phase and washed to remove the excess of the monomer and surface-active compound.

The permeability of the membrane is an important parameter of microcapsules. In principal, the membrane in each particular case can be permeable, impermeable or semipermeable. Impermeable membranes are not used for the encapsulation of therapeutic enzymes. Completely permeable membrane, i.e. a membrane with pores, the size of which permits free penetration of an encapsulated matter, can be used for the preparation of controlled release systems and drug depots in the organism. The release rate of an encapsulated compound can differ in rather wide limits by varying the membrane thickness. The same result can be achieved when a microcapsule is made from material capable of gradual biodegradation in physiological conditions or gradual swelling accompanied with increase in pore size.

Most studies are devoted to enzyme immobilization in semipermeable microcapsules, i.e. microcapsules made of polymeric material with a pore size sufficient for free penetration of low molecular weight substrates and products of enzymatic reaction but insufficient for high molecular weight proteins. Usually, the term "artificial cells" is applied to microcapsules of the last type, which have a relatively thin membrane (ca. 0.2 μm) which permits fast diffusion of low molecular weight soluble compounds from and into the microcapsule.

Mechanical properties of microcapsules should be taken into account when microencapsulated enzymes have to be used in column reactors or extracorporeal shunts. The mechanical properties are also very important in imparting the ability to pass through the capillary network.

Rather often, membranes of "artificial cells" are formed from biological or biodegradable materials: cross-linked or polymer-adsorbed proteins, lipids, lipid-polymer or lipid-protein complexes and polysaccharides. Proteins can be also included into membranes by physical adsorption on the surface of preformed microcapsules. As a result, the latter can improve their biocompatibility. Nonthrombogenic "artificial cells" were prepared by coating microcapsules with heparin [191].

Depending on the designation and microcapsule properties, microencapsulated enzymes can express their activity in the organism in several different ways. Thus, an enzyme can perform its function either by remaining inside the capsule or by leaving it. Consequently, different modes of administration of microencapsulated enzymes into the body can be suggested: intravenous, intramuscular, intraperitoneal, subcutaneous, peroral and local (as applications on wounds, ulcers and burns).

As already mentioned, the treatment of enzymatic digestion insufficiency is a very important task of clinical gastroenterology. This insufficiency can arise as a result of pancreatic diseases or surgery of digestive organs. Preparations of microencapsulated digestive enzymes are promising. It seems rational to include individual digestive enzymes or their mixtures into polymeric microcapsules capable of dissolution in the acidic medium of the stomach or in the alkaline medium of the intestine [208], because after peroral administration, native digestive enzymes are rapidly inactivated in the gastric juices. Pancreatic enzymes have been encapsulated into membranes made of acetylphtalylcellulose, which is not soluble in the gastric juice, but easily dissolves in the intestinal medium [196, 208].

Usually, an appropriate oily filler is introduced into microcapsules together with enzymes, because in the absence of this filler some components of the gastric juice can penetrate the membrane and inactivate the enzymes (some data on the encapsulation of a mixture of pancreatic enzymes, the so-called pancreatin, are presented in Fig. 24).

Another preparation for enzyme replacement therapy is the microencapsulated yeast invertase [209]. This preparation is necessary for the treatment of patients suffering from inherited sucrose intolerance, caused by lack of sufficient invertase activity in the intestinal mucosa. Therapy with the native enzyme requires very high doses of the protein because of its fast inactivation in the stomach at low pH values. The use of microencapsulated invertase allows to drastically decrease the enzyme quantity.

Polypeptide secretin, entrapped into microcapsules made of pH-sensitive polyacryloyl lysine, has been used for regulation of pancreatic secretion [210]. In this case, the release rate of secretin increases when the pH decreases, which permits to stimulate pancreatic juice secretion in the duodenum containing acidic chyme from the stomach. Sodium bicarbonates present in the pancreatic juice neutralize excessive acidity, which results in the inhibition of secretin release and pancreatic stimulation (biological feedback mechanism).

Microencapsulated urease has been suggested for peroral administration [193, 211]. In this particular case, the ammonia adsorbent has to be encapsulated together with urease, which absorbs the gas released in the urease-catalyzed reaction. Experiments on dogs have been performed, using animals pretreated with chemicals suppressing the growth of intestinal urease-producing bacteria. Microencapsulated urease successfully replaced the native enzyme and effectively converted urea into ammonia, the latter being absorbed by a coencapsulated adsorbent. The urea content in the blood was decreased by 40% in 4 hours. The

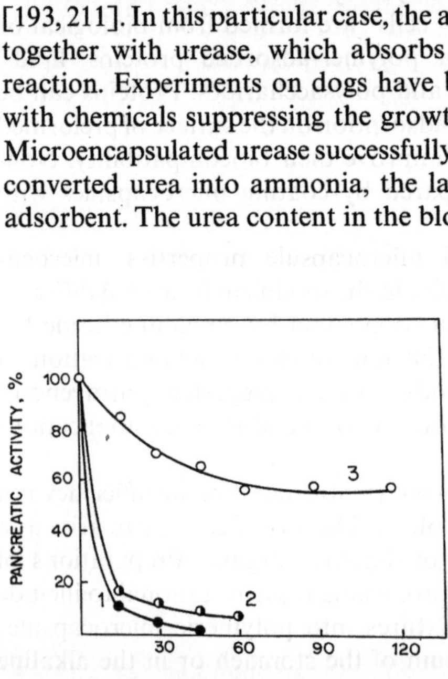

Fig. 24. Inactivation kinetics of native (*1*), microencapsulated (*2*) and microencapsulated together with a filler (*3*) pancreatin in gastric juice, pH 1.2, 37 °C [196]

authors of [212] suggested in order to increase the stability and efficacy of the preparation, preliminary immobilization of urease on silica with subsequent glutaraldehyde treatment and microencapsulation into acetalbutyrylcellulose membranes. Here, the ammonia adsorbent was used as the carrier for enzyme immobilization. Another peroral preparation is described in [213] where urease was encapsulated into cellulose acetatebutyrate together with zirconium phosphate as ammonia adsorbent. Experiments in dogs revealed the high efficacy in uremia treatment. Clinical studies performed on 6 patients also demonstrated the positive effect of the immobilized enzyme preparation.

The first model for the practical application of "artificial cells" was the use of microencapsulated urease to decrease urea levels in the blood of animals with experimental uremia [191, 200]. It was shown, that after intramuscular administration, microencapsulated urease effectively decreased urea levels by conversion into ammonia.

Hemoglobin-stabilized phenylalanine ammonia lyase has been included into cellulose nitrate microcapsules [236, 237]. The enzyme is used for replacement treatment in phenylketonuria. Microencapsulation of the enzyme does not change its catalytic properties, but noticeably increases its stability under physiological conditions. As a result, oral administration of the microencapsulated enzyme over a 2-day treatment course in mice with phenylketonuria (blood phenylalanine level being by 15–20 times higher than normal) decreased the content of phenylalanine in the blood of mice by 35% and over a 7-day course by 75%.

Many studies are devoted to the production of microencapsulated enzymes for parenteral administration. For this purpose, many therapeutic enzymes were included into semipermeable polymeric microcapsules—urease [193, 200, 211–217], catalase [214, 218, 219], asparaginase [201, 220–223], uricase [191, 194, 200], arginase [225], tyrosinase [226], histidase [227], desoxyribonuclease [228], carbonic anhydrase [191, 200, 229], trypsin [191, 199, 200, 230], chymotrypsin [202, 231], alkaline phosphatase [196], β-galactosidase [204, 232, 233], lipase [214], glucose oxidase [198], α-amylase [234], pronase [235], phenolase and nitrate reductase [32] and others.

As already mentioned, the stability of the enzymes listed above in microcapsules can be additionally increased by the coencapsulation of different fillers and by pretreatment of proteins with cross-linking agents or their preliminary immobilization on soluble polymers, or even by entrapment into microparticles of a polymeric gel. Thus, it was shown [221] that asparaginase pretreated with glutaric dialdehyde, in the presence of hemoglobin and than included into microcapsules, preserves up to 20% of its initial activity after incubation for 120 days at 37 °C, whereas the native enzyme completely loses its activity already in 3 days (see Fig. 25).

Insufficient catalase activity in man and animals causes acatalasemia. In acatalasemic mice, the activity of catalase in the blood and in the body reaches only two to 20% of normal levels. In these animals intraperitoneal administration of microencapsulated catalase successfully replaces the deficient enzyme [191,

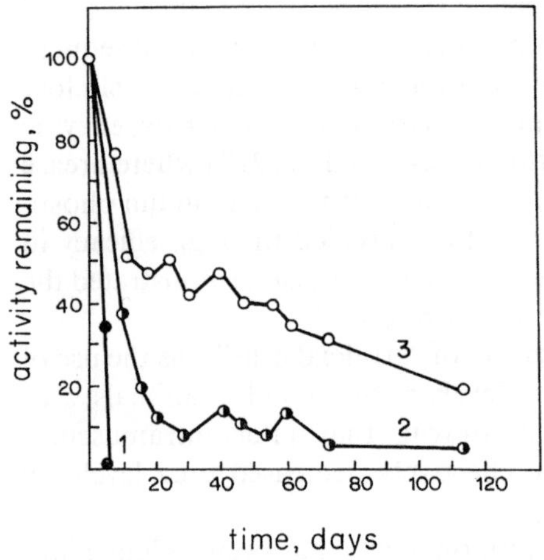

Fig. 25. Stability of asparaginase in solution at 37 °C (10 IUml). (*1*)—native enzyme, (*2*)—enzyme microencapsulated with 10% hemoglobin, (*2*)-enzyme microencapsulated with 10% hemoglobin and cross-linked with glutaraldehyde [221]

218, 219]. It has been found that the injection of a 50% (w/v) aqueous suspension of microencapsulated catalase (0.5 mmol per g of body weight) increased blood catalase activity in mice by 50% in 20 min after preliminary administration of a substrate (perborate), without any undesirable side effects. Microencapsulated catalase has also been used for local wound healing in man [238]. Acatalasemia causes the appearance of wounds and ulcers in the oral cavity of man and animals, due to the presence of free hydrogen peroxides produced by some bacteria, which belong to the normal microflora of the oral cavity. The native catalase which is used for treatment of these injuries in a water solution causes pronounced allergic reactions. However, local administration of the microencapsulated enzyme leads to effective degradation of hydrogen peroxide and wound healing without any immune response.

A great deal of effort has been devoted to obtain microencapsulated asparaginase, which is being used in the treatment of asparagine-dependent tumors. A number of studies [193, 222, 223] indicates the possibility of effective suppression of tumor growth with a microencapsulated enzyme.

Native asparaginase, when administered intraperitonealy into mice with implanted lymphosarcoma cells, caused a stable and prolonged decrease in the asparagine levels in the blood which inhibited tumor growth [193]. The time of tumor development in mice was 7–10 days without enzyme treatment and 10–18 days when native asparaginase was administered. At the same time 50 per cent of mice which obtained a similar dose of microencapsulated asparaginase did not show any tumor development during 120 days of observation.

Microencapsulated arginase also demonstrates antitumor activity in the treatment of L1210 leukcemic mice [225]. The therapeutic effect is based on the decrease of arginine concentration in the blood which results in the inhibition of arginine-dependent tumor growth. Data are available on the treatment of histidemia with microencapsulated histidase [227], which caused strong and prolonged decrease of the histidine level in the blood of experimental animals.

An interesting approach is the use of microencapsulated enzymes and proteins to substitute the oxygen-transporting function of red blood cells. Hemoglobin [191, 239], red blood cell hemolyzate [191, 240] and red blood cell hemolyzate mixed with catalase and carbonic anhydrase [191, 193] have been used for this purpose in animal experiments. Microencapsulated hemoglobin preserved its ability to bind, to transport and to release oxygen. The presence of catalase noticeably decreased the formation of methemoglobin in the hemolyzate; carbonic anhydrase decomposed carbon dioxide.

In the majority of the cases described, the active enzyme remained inside the microcapsule. It is evident that when microcapsules are made of synthetic polymers, repeated injections of microencapsulated enzymes can overload the organism with polymeric material. Two principal ways to solve this problem exist. Either, microencapsulated enzymes can be used as components of extracorporeal shunts, or, microcapsules are made of biodegradable materials (cross-linked proteins, protein-polymer composites, lipid-protein mixtures, polylactic acid, polyalginate). At present, most popular material is polylactic acid which easily metabolizes in the body. Microcapsules made of polylactic acid have been already used for the immobilization of different physiologically active compounds and have demonstrated good properties in animal experiments [241–243]. The most interesting results were obtained with asparaginase, trypsin, insulin and trypsin pancreatic inhibitor immobilized into polylactic acid microcapsules [244]. It was shown that the rate of polylactic acid microcapsule biodegradation under physiological conditions can be controlled by the incorporation of different charged substances into the microcapsules. Thus, the incorporation of tertiary amine increases the rate of microcapsule biodegradation [245].

A relatively new method is the encapsulation of drugs into microcapsules made of polyelectrolytes [245a]. Rat liver microsomes have been entrapped in capsules prepared from polyelectrolytes. Despite of the lowered activity of encapsulated microsomes (by 75%) (deethylase activity of microsomes against p-nitrophenetole) established by determination of this method of enzyme immobilization may be promising because of the mild immobilization conditions.

Microcapsules can be also used for the immobilization of clinically significant multienzyme systems [251, 252]. These systems can be used for cofactor regeneration [251] and for the removal of toxic metabolites, such as urea. In the latter case, according to the scheme presented in Fig. 26, urea and ammonia can be converted into aminoacids by a system consisting of three co-encapsulated enzymes [252].

Successful experiments with encapsulation of pancreatic Langerhance islets (246, 247), (included into microcapsules made of non-toxic sodium alginate additionally treated with polylysine) must be mentioned here. In vitro, microencapsulated islets remain intact and functionally active for 15 weeks. The membrane is permeable for glucose and insulin molecules. The use of microcapsules in mice with experimental diabetes demonstrated that intra-

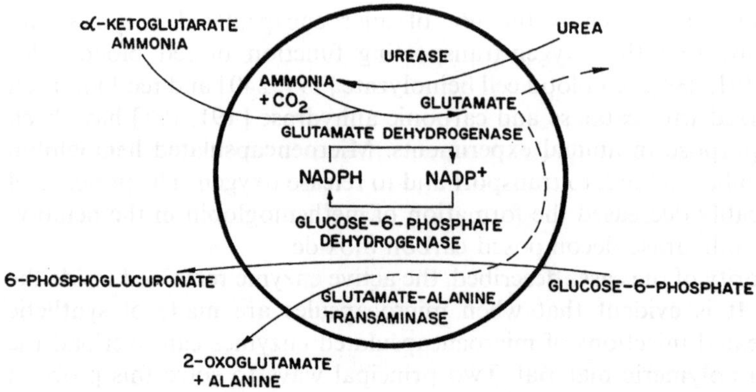

Fig. 26. Multienzyme system to convert urea and ammonia to simple amino acids. Schematic representation of a multienzyme system for the conversion of toxic waste metabolites into harmless metabolic by-products (Reproduced with permission from Poznansky MJ, Juliano RL (1984) Pharm Rev 36:277)

peritoneal administration causes decrease in blood glucose which is much more stable and prolonged than that after microencapsulated insulin administration.

Numerous additional examples on microcapsule preparations (from polyamides, polylactides, ethylcellulose etc.) and their use for the immobilization of different enzymes and drugs can be found in [248–250].

4.1.2.2 Properties and Application of Enzymes Immobilized in Liposomes

During the past 15 years another method of immobilization—encapsulation in liposomes—has been widely studied. Liposomes are artificial phospholipid vesicles, obtained by various methods from lipid dispersions in water. The problems connected with liposome preparation, their physico-chemical properties and possible biomedical application have already been discussed in several monographs [253–257].

As drug (enzyme) carriers, liposomes possess a whole set of evident advantages. Thus, liposomes are suitable for the immobilization of various substances—water-soluble compounds are entrapped into the liposomal inner water space and the water space between bilayers (in case of multilammellar liposomes), and lipophilic compounds may be incorporated into the phospholipid membrane (see the scheme in Fig. 27).

From the biomedical point of view, liposomes appear to be biocompatible, causing no antigenic, pyrogenic, allergic and toxic reactions; they easily undergo biodegradation; they do afford host protection from any undesirable effects of the encapsulated enzyme, at the same time protecting an entrapped enzyme from the inactivating action of the physiological medium; and, last but not least, liposomes are capable to deliver their content inside many cells (the principal mechanisms of liposome-cell interaction are presented in Fig. 28). Recently, a further method of

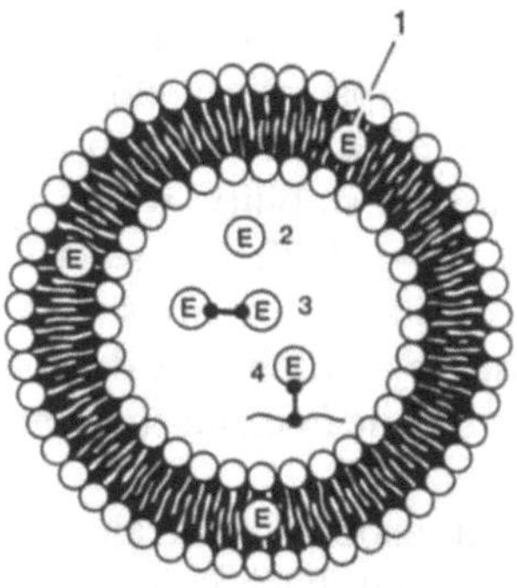

Fig. 27. Enzyme immobilization into liposomes. Enzyme (*E*) entrapment into liposomes: hydrophobic enzymes are included into the phospholipid membrane (*1*); hydrophilic enzymes can be immobilized in the native (*2*) or stabilized form (*3*-intermolecularly cross-linked, *4*-bound with soluble carrier) (Reproduced with permission from Torchilin VP (1987) Adv Drug Deliv Rev 1:41)

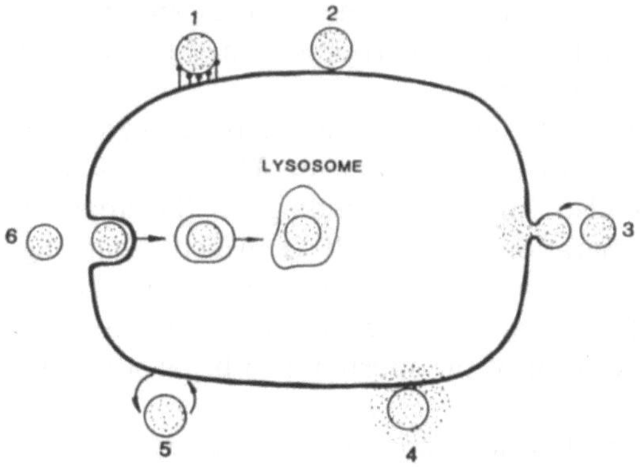

Fig. 28. Liposome interaction with cells. Liposomes can interact with a cell via specific adsorption (*1*), non-specific adsorption (*2*), fusion (*3*), release of an entrapped matter in the vicinity of cell membrane with its subsequent penetration into the cell (*4*), lipid exchange with the cell membrane (*5*) and endocytosis (*6*). (Reproduced with permission from Torchilin VP (1987) Adv Drug Deliv Rev 1:41)

liposomal content delivery into the cytoplasm has been elaborated [258]. According to this method, the liposome is made of pH-sensitive components and, after being endocytosed in the intact form, it fuses with the endovacuolar membrane under the action of lowered pH inside the endosome, releasing its content into the cytoplasm. The method described seems to be very promising for intracellular drug delivery, which opens unique opportunities in the enzyme therapy of diseases caused by inherited disturbances in the normal functioning of intracellular enzymes, e.g., in liver cells [5, 6, 259].

By now, many different methods have been suggested to prepare liposomes of different sizes, structure and size distribution. A detailed description of possible methods of liposome preparation is given in Ref. 260, albeit since then, a number of original techniques has been suggested e.g., [261, 262]. The most frequently used methods are ultrasonication, reverse phase evaporation and detergent removal from mixed lipid-detergent micelles by dialysis or gel-filtration. These methods, together with several more exotic ones, permit now to obtain liposome preparations with desired properties and very high encapsulation efficacy. Besides, during the last several years, methods were developed to obtain lyophilized liposomal preparations possessing good storage stability, which is very important from a practical point of view [263, 264]. Different sugars, including threhalose, turned out to be good cryoprotectants for liposomes. Lyophilized liposome preparations obtained in the presence of threhalose do not lose more than a few per cent of encapsulated drug during one freezing-thawing cycle.

Various phospholipids have been used to obtain liposomes: neutral, e.g., natural phosphatidylcholine (lecithin) or synthetic phosphatidylcholine dipalmitoyl and phosphatidylcholine dimyristoyl; negatively charged, e.g., phosphatidic acid or dicetylphosphate; and positively charged, such as phosphatidylethanolamine. To increase liposome stability towards the action of an agressive physiological environment, cholesterol can be incorporated in the liposomal membrane (sometimes up to 50% mol) [265].

The size of liposomes depends on their composition and preparation method and can vary from 25 nm to greater than 1 μm in diameter. The encapsulation efficacy for different substances is also variable depending on the liposome composition, size, charge and preparation method. It is evident that, e.g., negatively charged substances will be better encapsulated in positively charged liposomes and vice versa. The use of the reverse phase evaporation method [260] permits inclusion of 50 and more per cent of the substance to be encapsulated from the water phase into the liposomes.

The in vitro release rate of different compounds from liposomes, including proteins of moderate molecular weight, such as lysozymes or insulin, is usually ~1% per hour, under the condition that the incubation temperature sufficiently differs from the phase transition temperature of a given phospholipid. Maximal permeability of liposomes is usually observed at temperatures close to the phase transition temperature of liposomal phospholipid. In vivo, this parameter can vary within wide limits—from minutes to hours—and depends on the liposomal membrane composition, cholesterol content and liposome location in the body.

Biodistribution of liposomes is a very important parameter from the clinical point of view. Liposome accumulation in different organs and tissues has been investigated in numerous animal experiments, the results of which have been summarized [254–256, 266–268]. Many studies have shown that within the first 15–30 min after intravenous administration of liposomes, between 50 and 80% of the dose is adsorbed by the cells of the reticuloendothelial system, primarily by the Kupffer cells of the liver. Considerable amounts of liposomes are also found in the spleen and in the lungs, and part of the liposomes and their degradation products are either found in the bone marrow or in the circulation. When subcutaneous, intramuscular or intraperitoneal administration is used, the general picture of liposome distribution in the organism changes, despite that sooner or later, the major part of the liposomal material becomes associated with the liver. The most controversial data are obtained in the case of peroral administration of liposomes [269, 270]. The method was developed almost exclusively for diabetes therapy and oral insulin administration, but if workable could be used in many other situations, including enzyme therapy. Unfortunately, at present one can not be certain about the fate of perorally administered liposomes and about the ability of the liposome-incapsulated drug to reach the circulation from the gut or intestine through many cellular and membrane barriers.

For the practical clinical use of liposomes, it is very important to know the exact mechanisms of liposome interaction with different cells and blood components. Numerous in vitro experiments elucidate the transfer into cell of liposome encapsulated materials (enzymes mainly, [271–272]). The fusion of liposomes with cells, resulting in the delivery of intraliposomal material into the cell cytoplasm, is a relatively rare event and requires special efforts for its realization [258]. Endocytosis of liposomes by cells usually leads to the delivery of liposome-encapsulated matter into lysosomes, where a drug can be degraded before it can perform its therapeutic function [273, 274]. Despite the fact that after endocytosis, the liposomal content can penetrate endovacuolar membranes and enter the cell cytoplasm (see, e.g., Ref. 275), the development of reproducible methods of cytoplasmic drug delivery by means of liposomes is still a matter of the future.

Detailed studies of liposome interaction with cells and blood plasma components have been performed [273, 274] with the following important conclusions: first, liposomes, especially non-stabilized ones, degrade relatively fast in the blood and lose their content, primarily because of their interaction with high density lipoproteins. Second, a large part of liposomes is captured by liver cells before liposome degradation in the blood. Third, the capture of liposomal material by hepatocytes proceeds either indirectly, via high density lipoproteins, or directly. The latter can be observed for small liposomes capable of penetrating the endothelial barrier. These conclusions were supported by many other investigators [276–279].

The first enzyme—lysozyme—was incapsulated into liposomes made of phosphatidyl choline, dicetyl phosphate and cholesterol in 1970 [280]. To date, several dozends of enzymes have been encapsulated into liposomes, including glucose oxidase [281]; glucose-6-phosphate dehydrogenase, hexokinase and β-galactosidase [282, 283]; β-glucuronidase [284]; glucocerebrosidase [285]; α-mannosidase [286], amiloglucosidase [287, 288]; hexoseaminidase A [289]; peroxidase [290–292]; β-D-fructofuranosidase [293–295]; neuraminidase [296, 297]; superoxide dismutase and catalase [298]; asparaginase [299, 300]; cytochrome oxidase [301]; ATPase [302]; dextranase [303, 304] and many other enzymes from different sources, such as plants, bacteria and animals. In accordance with the recommendations of [272] the following parameters are proof of enzyme incorporation into the inner water phase of liposomes: the possibility of chromatographic separation of liposome-encapsulated and free enzyme; the latency of liposome-encapsulated enzymes; the correlation between protein incorporation and an increase in net surface charge of the lipid bilayer.

Enzymes can be immobilized into liposomes without affecting enzyme specific properties; entrapment efficiency can also be rather high (see data concerning some hydrolases in Table 11).

From the clinical point of view, the ability of intraliposomal enzymes to enter the cytoplasm or lysosomes is of primary importance for the treatment of inherited diseases, connected with the abnormal functioning of some intracellular enzymes. Despite relative rarity of these diseases, together they pose a serious medical problem [305, 306].

Storage diseases can be divided into two main groups [272]. In the first group, an abnormal enzyme is localized in the vacuolar apparatus of the cell. As a result, non-degraded substrates accumulate in the lysosomes of affected cells. Diseases and as Tay-Sachs', Gaucher's, peroxidase unsufficiency and different mucopolysaccharidoses are typical examples. The second group includes cases, when the activity of cytoplasmic enzymes is decreased or in some was deteriorated: here, non-degraded substrates accumulate in the cytoplasm. Pathologies such as the Lesh-Nyhan syndrome or adenosine desaminidase insufficiency are representatives of this group.

Unfortunately, the attempts to treat these diseases by simple infusion of the deficient enzyme were unsuccessful. Intravenous administration of purified hexoseaminidase has been tried in the treatment of Tay-Sachs' disease [307]. The enzyme was thought to enter the affected cells of the central nervous system via endocytosis and to destroy the ganglioside GM_2 accumulated in cell lisosomes. It was found that endocytosis can not provide sufficient amounts of the enzyme inside the cells; furthermore, predominant accumulation of hexoseaminidase was

Table 11. Entrapment of β-glucosidase and other hydrolases in liposomes. Enzyme-rich material (2 ml), as obtained in the last step of the purification procedure and containing approx. 8800 units of N-acetyl-β-glycosaminidase, 3600 units of α-mannosidase, 6300 units of α-glucosidase, 1100 units of β-galactosidase and 4500 units of α-glucosidase were entrapped in liposomes composed of egg phosphatidylcholine (30.0 mg) and cholesterol alone or supplemented with charged lipids. Entrapped enzyme activity (5% of activity used) was more than 95% latent and it could only be measured on the addition of Triton X-100 (0.8% final concn.). (From Braidman I.P., and Gregoriadis G., Biochem. J. 164, 439, 1977, with permission)

Enzyme	Charged lipid	Molar ratio of lipids	Liposomal surface charge	Entrapment (%)
β-Glucosidase	None	7:2	Nil	100.0
	Phosphatidic acid	7:2:1	Negative	85.0
	Phosphatidic acid	7:2:2	Negative	65.0
	Phosphatidic acid	7:2:4	Negative	80.0
	Dicetyl phosphate	7:2:1	Negative	25.0
	Stearylamine	7:2:1	Positive	42.0
N-Acetyl-β-glucosaminidase	None	7:2	Nil	19.0
α-Mannosidase	None	7:2	Nil	32.0
α-Glucosidase	None	7:2	Nil	32.0
β-Galactosidase	None	7:2	Nil	99.0

discovered in the liver instead of CNS cells. Similar problems are linked to unsuccessful attempts with purified glucocerebrosidase and ceramidetrihexosidase [308, 309].

The use of liposome-immobilized enzymes instead of their native precursors opens new opportunities for enzyme therapy [310, 311]. Thus, the biodistribution of liposomes made of phosphatidylcholine, phosphatidic acid and cholesterol (7:1:2 molar ratio) and containing β-fructofuranosidase has been studied [293]. It was shown that within an hour, ~50% of the administered enzyme can still be found in the circulation, but after 6 hours, 45% of the enzyme activity accumulated in the liver. The enzyme preserves its activity for a long time—25% of the administered activity can be found in the liver after 48 hours. Even after 100 hours, 5% of active enzyme are still present. It is very important that up to 50% of intracellular enzyme activity is localized in the lysosomal fraction. Similar data have been obtained for intravenously administered liposome-encapsulated α-mannosidase [286] and neuraminidase [296].

β-Glucuronidase, immobilized into charged liposomes, composed mainly from phosphatidyl choline dipalmitoyl, also demonstrated fast accumulation in the liver of experimental mice. The enzyme remained active more than a week, associated with the lysosomes of liver cells [284].

After lysosome overloading with dextran administered intravenously into mice, the dextran content in the liver of control mice remained unchanged for 6 days, whereas rats treated with liposome-immobilized dextranase (intravenous injection) demonstrated the decrease in the dextran content by 70% in 2 days [303, 304].

Glucocerebroside β-glucosidase of human origin was encapsulated into neutral liposomes made of egg lecithin [312]. The efficiency of enzyme entrapment into liposomes is very high, the latency of the encapsulated enzyme is more than 95%. The ability of liposome-immobilized β-galactosidase to degrade GM_1-ganglioside in lysosomes of feline fibroblasts with pathological accumulation of this substrate, has been demonstrated [313]. The native enzyme was unable to penetrate cells and to perform the therapeutic function.

Later it was shown that the incorporation of liposome-encapsulated therapeutic enzymes into appropriate cells can be enhanced by rather simple methods [314]. Thus, β-glucocerebrosidase included into liposomes was predominantly captured by Kupffer cells in the liver; modification of liposomes with mannoside residues increased the capture.

It follows from the data presented that the liver is the natural target for liposomes. The earliest attempts to use liposomes as drug carriers were connected with the treatment of diseases localized in cells of reticulo-endothelial system. The first success in this field was achieved in experimental therapy of leishmaniasis [315, 316] caused by protozoa, which localize in the cells of the reticulo-endothelial system. The encapsulation of toxic drugs (arsenic or antimony derivatives) into liposomes leads to their spontaneous accumulation in infected organs. According to different estimates, the efficiency of the liposomal drug is 700 to 1 000-times higher than that of routine preparations.

The first clinical experiments utilizing liposome-encapsulated therapeutic enzymes have been also performed with preparations designed for the treatment of disease of the reticulo-endothelial system. Impressive results have been obtained by treatment of a patient suffering from Gaucher's disease (inherited glucocerebrosidase insufficiency, causing excessive glucocerebroside accumulation in liver lysosomes), using liposomal glucocerebrosidase [285, 317]. Here, therapy with the native enzyme gave no results because of its inability to penetrate cells. Administration of enzymes, encapsulated in liposomes, over a period of 13 months produced a pronounced effect, in particular a decrease in liver size with a very small amount of the active enzyme used. An attempt to treat a patient suffering from type II glycogenosis or Pompe's disease, is also somewhat encouraging. Here, 3 mg of amyloglucosidase immobilized in liposomes (170 mg phospholipids, phosphatidylcholine, phosphatidic acid and cholesterol in 7:1:2 molar ratio) was injected intravenously over a period of 1 week into an affected child, which led to a measurable decrease in liver size, though later the patient died. The liver still enhibited glucosidase activity post-mortem.

Although in a number of studies, liposomes have been shown to accumulate in various organs and tissues, particularly in those exhibiting pathological lesions (e.g., accumulation of neutral and positivey charged liposomes in zones of experimental myocardial infarction and in cardiomyocytes in culture [318, 319]), the approach based on spontaneous liposome accumulation can hardly be considered as universal for the delivery of liposomal drugs (enzymes) in different organs. Liposome circulation half-life needs to be prolonged and their capture by the cells of the reticulo-endothelial system decreased. For the former problem, methods for long circulating liposomes are being investigated. For a long time, it was assumed that different phospholipases (phospholipid acyl hydrolases and phospholipid acyl transferases) are responsible for liposome degradation in the circulation. Having this in mind, numerous attempts have been made to obtain liposomes from synthetic phospholipid analogues which are not substrates for these enzymes, because of some chemical rearrangements in their phospholipid molecules. The results obtained are not completely clear yet. Liposomes have been described with a half-life in the circulation of about 100 hours; on the other hand—it was shown that in some cases, liposome degradation products and not intact liposomes are present in the blood in the case of "modified" liposomes [320, 321]. Besides, synthetic phospholipid analogues are usually more toxic than natural phospholipids.

To prevent liposome capture by the liver, it was suggested to perform a preliminary blockade or saturation of hypothetic liposome-binding areas in the liver, using inert particles such as "empty" liposomes. In this case, liposomes containing an encapsulated drug should be administered some time after the injection of inert particles. Animal experiments supported this idea and showed that the circulation half-life of therapeutic liposomes can be somewhat increased [287, 295]. It was also suggested to "coat" the liposome surface with some proteins or aggregates (albumin, globulins, lipoproteins) which are normally present in the blood in high concentrations and saturate appropriate binding sites in the liver [322, 323]. Stable liposomes can be also prepared from so-called "polymerizable" phospholipids [324]. To obtain such liposomes, unsaturated groups are chemically introduced into phospholipid molecules. As a result liposomes acquire the ability to polymerize following the action of different initiators. The polymeric network, formed in the liposomal bilayer after polymerization, additionally stabilizes it against different factors.

Animal experiments have also clearly demonstrated the suitability of liposomes for immobilization of enzymes used for the therapy of diseases not located in the liver. Thus, the increase in circulation half-life of liposomal L-asparaginase (in lecithine-dicetylphosphate liposomes, a negative charge is necessary to inhibit liposome interaction with cells) and the decrease in its antigenicity and susceptibility towards proteolytic degradation together with the increase in efficacy of experimental tumor therapy in mice have been shown [300].

The authors of [325, 326] described the original use of liposomal drugs in cancer therapy. They suggested the destruction of malignant cells by hyper-activated macrophages of the host himself. For this purpose, macrophages have been stimulated with different lymphokines. In order to increase the quantity of lymphokines delivered to macrophages, the former can be preliminarily encapsulated into liposomes. In experiments on mice with inoculated tumors it was shown that liposomal lymphokines are capable of preventing metastasis formation in 70% of the animals, in comparison with a 10% prevention when native lymphokines were used.

Liposomes can be also used for the transmembrane intracellular delivery of superoxide dismutase and catalase [298, 327, 328]. This is extremely important for the elimination of oxygen-derived free radicals inside cells, as their increased generation causes the toxic effect of ischemia on endothelial and many other cells. Using liposomes, intracellular superoxide dismutase activity can be increased by 15-fold.

The use of liposomes for the transfer of therapeutic enzymes through the "blood-brain" barrier which permits to deliver these enzymes into cells of the central nervous system also seems very attractive. It has been already shown that horse raddish peroxidase, encapsulated into liposomes made of phosphatidylcholine, cholesterol and phosphatidic acid (7:2:1 molar ratio), acquires the ability to cross the hemato-encephalic barrier, whereas the native enzyme can not. The presence of peroxidase in brain cells was proved by histochemical methods [328a]. The same authors have shown that after injection of liposomal glucose oxidase into the rat tail vein, up to 5% of the enzymatic activity can be discovered in brain tissues [329]. This opens interesting opportunities in the treatment of inherited enzyme storage diseases, localized in brain cells and in the delivery of different anti-tumor antibiotics used for the treatment of brain tumors [330, 331]. Leupeptide (the tripeptide inhibitor of proteolytic enzymes) can be delivered into the brain by means of liposomes obtained by reverse phase evaporation from a mixture of lecithine, cholesterol and sulfatide (4:5:1 molar ratio) [332]. The inhibitor was used for the treatment of experimental allergic encephalomyelitis in guinea pigs. The therapeutic efficiency of the liposomal inhibitor was very high as estimated by histopathology data and the survival of experimental animals. Liposome-encapsulated inhibitors of aldose reductase (quercitine, quercitrine, AU22-284 and sorbinyl) have been successfully delivered into the ocular lens [333–335].

β-Galactosidase-containing liposomes obtained from a mixture of lecithine, cholesterol and sulfatide (in 7:2:1 molar ratio) are readily incorporated into the liver and spleen of mice hosting the model of globoid cell leukodystrophy. The liposome-encapsulated therapeutic enzyme, administered as a single injection after preliminary injection of liposomes with galactocerebroside into experimental mice, causes the break-down of 70–80% of intracellular galactocerebroside [336].

The inherited abnormal metabolism of porphyrins results in porphyriasis. The molecular basis of the disease can be correlated with the increase in δ-aminolevulinate synthetase activity in conjunction with normal or even decreased activity of δ-aminolevulinate dehydratase. The authors of [337] succeeded to incorporate purified δ-aminolevulinate dehydratase into liposomes. Under optimal conditions, the encapsulation efficiency can reach 40% and the enzyme completely preserves its activity.

The efficiency of intracellular cGMP formation was increased by the treatment of cells (erythocytes) with liposomes, containing the membrane form of guanylate cyclase [338]. In this case, the enzyme was delivered intracellularly via liposome-cell fusion induced with polyethyleneglycol 4000.

From the practical point of view, new artificial oxygen-transporting systems capable of prolonged activity in the circulation are of special interest. The topic has been extensively elaborated; we will mention only several examples; where natural hemoglobin was incorporated into liposomes of different composition (so-called, hemosomes). Thus, it was shown that the maximal quantity of hemoglobin obtained from lyzed erythrocytes incorporates into negatively charged liposomes [339]. The authors of [340] have stabilized hemosomes with carboxymethylchitine. Stabilized hemosomes bind oxygen in the same way as human blood hemolysates. The acute toxicity of hemosomes was moderate—in mice, the LD_{50} was 13.8 ml hemosomes per kg weight. Polymerizable liposomes [341, 342] have also been used for hemosome preparation. Stable polymerized hemosomes are capable of reversible binding of molecular oxygen in physiological conditions even at high flow rates [341]. The incorporation of allosteric effectors into hemosomes permits almost quantitative conversion of immobilized hemoglobin into the oxy-form [342].

Liposomes have been also used for the immobilization of some other non-enzymatic proteins, which are of interest from the clinical point of view. The authors of [343] have described a preparation of liposomes containing blood clotting Factor VIII, which can be used for the treatment of hemophilia. Experiments have been done on the peroral administration of Factor VIII containing liposomes into hemophilic dogs. Angiotensin II [344] and heparin [345] have been also included into liposomes. An interesting attempt has been made to treat viral diseases with the liposome-encapsulated fragment A of diphteria toxin [346]. It has been shown in cell culture experiments (cells infected with the virus of sclerosing panencephalitis), that the liposome immobilized A fragment decreases cell infection by 99%, whereas the free fragment itself does not influence infected cells at all.

Many studies deal with peroxidase incorporation into liposomes and the delivery of liposome immobilized enzyme into different organs and tissues [347–349]. In particular, liposomal peroxidase was incorporated into the nerve-endings of the ray electric organ. Liposome immobilized cytochrome oxidase from bovine heart was delivered into rabbit erythrocytes [350]. Urease [351] and glutathione transferase [325] have been also obtained in liposomal form. Liposomal DNAase I can provoke neoplastic transformation in embryonal Syrian hamster cells in culture [353]. It was also shown that the liposomal membrane can influence catalytic properties of liposome immobilized β-D-galactoside 2→3 sialyltransferase [354]. The probability of similar effects should always be taken into account with liposome immobilized enzymes (especially, membrane enzymes).

In some cases proteins, enzymes among them, have been immobilized on the outer surface of liposomes. Insulin, immobilized on the outer surface of multilamellar liposomes via SPDP, preserves its antigenic specificity [355]. Comparative studies of different methods for the immobilization of asparaginase and peroxidase on liposome surfaces have revealed that in both cases the maximal quantity of a protein with maximal activity preservation can be immobilized by preliminary modification of the proteins with hydrophobic groups [356].

Special attention has been paid to the immunological properties of liposome immobilized enzymes. The ability of liposomes to demonstrate in some cases adjuvant properties is already well known [357]. The authors of [358] demonstrated the possibility of enhanced immune response development in mice receiving intravenously liposomes with immobilized bovine β-glucuronidase, despite the expected protection of the intraliposomal enzyme from contact with immunocompetent cells [358]. A pronounced immunoadjuvant effect of liposomes can also be seen when enzymes or other immunogens are bound to the outer surface of liposomal membranes [359].

Recently, special attention is paid to the investigation of side-effects, accompanying liposome administration. It has been shown that repeated liposome administration can provoke pathological changes in the mouse liver [360].

To predict the behaviour of liposomal drugs in the body, the pharmacokinetics of liposomal preparations has been widely investigated. It has been found, that the pharmacokinetics of liposome encapsulated drug depend to a great extent on liposome size and the release rate of encapsulated drugs from liposomes [361, 362].

Among new, nontraditional forms of liposomal preparation, one can find liposomes included into polymeric gel [363] and liposomes containing polymeric gel inside the vesicle [364]. The approach can be useful for controlled release systems [364, 365]. Some other methods of regulating membrane permeability have also been developed. Among them are the use of polyelectrolytes [366] or microwave irradiation [367].

4.1.2.3 Properties and Application of Enzymes Immobilized in Red Blood Cell Ghosts

At the beginning of the seventees a number of researchers suggested the use of cells (in particular, red blood cells), as drug carriers [368–370]. A mammalian erythrocyte is a highly specialized cell, designed for oxygen transfer, which does not contain organelles. A human erythrocyte has a concavo-concave disk shape with a diameter of 7–8 μm and a thickness of 1.5–2.0 μm. It has been demonstrated [371–373] that erythrocytes of different origin can be "opened", "loaded" with both low molecular weight or/and high molecular weight substances, "sealed' and returned into the blood flow. An erythrocyte possesses a number of advantages as a drug carrier. This includes it biocompatibility, biodegradability and the ability for prolonged circulation in the blood flow. A natural erythrocyte "lives" in the organism for approximately 120 days. Even when "loaded" with a drug, an erythrocyte is capable to circulate in the blood for 10–20 days [371–374]. One of the assets of an erythrocyte is its high internal volume—about 90 μm^3. On the other hand, the comparatively large size an erythrocyte can be considered as a drawback, since it is not possible for an erythrocyte to penetrate tissues to reach extravascular targets and thus the possibility of its application is limited to a radius of targets accessible from the blood flow. Another drawback of an erythrocyte, compared to other carriers, is the impossibility to vary its properties. Data on the application of erythrocytes (and other cells) as carriers are discussed in many reviews and monographs [375–378].

Methods of erythrocyte loading are based on the reversible formation of holes (pores) in the membrane through which the exchange between the internal contents of erythrocytes and the external medium proceeds. These methods can be divided into two groups, according to the type of the treatment used for pore formation: the application of hypotonic lysis and application of electroporation of the membrane [379].

In the case of hypotonic lysis, (see Fig. 29) the erythrocytes are placed into a solution, the osmotic pressure of which is lower than that of the erythrocyte content. Due to this, the internal volume of the cell begins to expand. Since the concavo-concave shape of the erythrocyte provides excess of surface in relation to volume, the erythrocyte first acquires a spherical shape. If at this moment the osmotic pressures inside and outside an erythrocyte do not become equal, further inflow of water leads to the formation of pores. Depending on the conditions, these pores can have different sizes—from dozens of Angströms to 1–2 μm [380–381]. After loading the erythrocytes with substances present in the solution (which usually takes several minutes), the osmolarity of the medium is restored to the initial level by addition of a concentrated saline solution and the erythrocyte

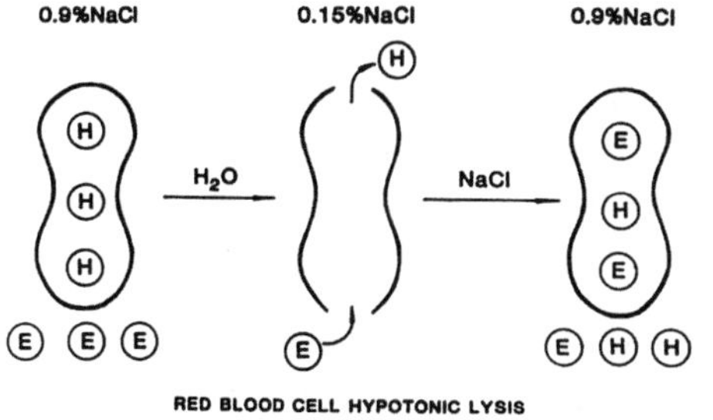

Fig. 29. Enzyme (E) entrapment into red blood cell ghosts by hypotonic lysis, *H*—hemoglobin (Reproduced with permission from Torchilin VP (1987) Adv Drug Deliv Rev 1:41)

suspension is incubated at 37 °C. This procedure leads to the closing of pores and restoration of membrane integrity. It is believed that even viral particles can be incorporated into erythrocytes by this method [379, 382].

Three types of hypotonic lysis can be defined: an osmotic shock, a preliminary swelling and a hypotonic dialysis [383]. The method of osmotic shock requires the addition of comparatively large volume of the substance to be incorporated into erythrocytes to the hypotonic solution. Hemolysis takes several seconds, and a minute later the equilibrium is established and the loading is completed. The method has two significant drawbacks: first, since for effective lysis it is necessary to add a considerable volume of hypotonic solution, the efficiency of loading will not be high; second, due to the same reason, the extent of loss of the contents by erythrocytes will be high. This can cause a significant reduction of life-span of cell "ghosts" in the blood flow. Nevertheless, there are reports that erythrocytes loaded by this method are capable to circulate in the blood flow for a long time [371, 372, 375, 384].

The method with preliminary swelling of cells consists the following procedures: erythrocytes are first incubated in the solution, the osmolarity of which is chosen to permit erythrocytes to swell singnificantly and the integrity of the membrane to be preserved. Then, erythrocytes are separated from the solution by centrifugation, and the cell precipitate is supplemented with a hypotonic solution of the substance to be loaded. At this stage, erythrocytes are opened and the drug can enter. Since erythrocytes are already on the verge of opening, the volume of the added solution can be small (about equal to the volume of cells). The following stages are the same as for normal hypotonic lysis procedure [385, 386].

The loading of erythrocytes using dialysis is performed as follow [386]. The erythrocyte suspension in a concentration of 70–80% vol. is dialyzed for 0.5–2 h against a hypotonic solution. At this stage, lysis of cells occurs. Then, the opened cells are taken off the dialysis tubing and the substance for loading is added. After incubation with this substance, the osmolarity of the system is restored and the pores are closed. Since in this method loading takes place practically without change of volume, the loss of high molecular weight components of the erythrocyte is minimal and the efficiency of incorporation of the loaded substance will be maximal [379, 382, 388].

A method of loading erythrocytes by electrical perforation of membrane has been suggested and developed [373, 376, 389, 390]. This method is based on the effect of the reversible formation of pores in biological membranes under the action of a transmembrane potential of 1 Volt order. It is performed experimentally in the following way: an erythrocyte suspension together with the substance to be incorporated is placed into a well with two electrodes at 4 °C, and electrical field with a tension of several kv/cm and a duration from several to tens of μsec is applied to these electrodes. After loading, the pores are closed by incubation of the "ghosts" at 37 °C. The isotonicity of the medium is preserved at all stages. The size of the pores depends on the tension of the field and duration of the impulse and it is possible to select optimal conditions for low molecular weight substances to enter erythrocytes without the loss of high molecular weight components of the cell [385, 386].

A number of enzymes (asparaginase, urease) can effectively function inside erythrocyte ghosts, since the membrane of erythrocytes is permeable for their substrates and reaction products. At the same time, heterologous enzymes appear to be protected from uptake by the liver, from contact with the components of the immune system, from the action of inhibitors and from filtration from the circulation comparable to the use of any other microcontainers [391, 393].

It is possible to induce the fusion of the bound carrier-erythrocytes with cultivated cells in vitro by polyethyleneglycol or by addition of inactivated viral particles and a short-term oxidation of the medium [383, 384].

Much attention has been paid to the development of simple and effective methods for the incorporation of proteins (enzymes) into erythrocytes. Thus, the authors of [394] have developed a dialysis-based method of immobilization which provides an efficiency of incorporation of up to 40%. This method has been tested with radiolabeled albumin and two enzymes—hexosaminidase B and β-glucosidase (see Table 12). The minimal efficiency of incorporation (with hexosaminidase) was equal to 36%; human erythrocytes from healthy donors were used as carriers. It has been found that the maximal incorporation is observed at hematocrit values of about 80%. The whole procedure takes 75 min at 4 °C. The loaded cells are able to circulate in the blood flow for 5–10 days.

Some low molecular weight drugs (cortisol, methotrexate) and also α_1-antiplasmin from plasma have been incorporated into rat erythrocytes using hypotonic swelling, which permitted to introduce the preparations into cells without lysis of the latter and with minimal impairement of the membrane [395]. The incorporation efficiency varied from 5 to 20%. Stability of the preparations obtained, as determined by release of hemoglobin, was close to that for normal cells.

The authors of [386] analyzed 7 methods of incorporation of a number of substances, inclucing β-glucuronidase, into human erythrocytes. Six of these methods were based on hypotonic action on cells, and one method on chlorpromazine-induced endocytosis. In all methods, the incorporation of low

Table 12. Incorporation of proteins into erythrocytes[a]. (From Dale G.L., Villacorte D.G. and Beutler E., Biochem. Med. 18, 220, 1977, with permission)

Protein	Concentration[b]	Incorporation (percentage)[c]
^{125}I-labeled human	0.1 mg/ml	38 ± 3(13)[d]
serum albumin	2.0 mg/ml	38
Hexosaminidase B	66 mU/ml	36
β-Glucosidase	29 mU/ml	42

[a] These entrapment experiment were done with 85% hematocritis. Lysis was carried out at 4 °C for 75 min against 25 to 100 vol of 5 mM KH_2PO_4/K_2HPO_4, pH 7.4. Resealing was for 30 min at room temperature versus 25 to 100 vol of 0.15 M NaCl, 5 mM KH_2PO_4/K_2HPO_4, pH 7.4;
[b] Initial concentration of protein or enzyme in dialysis bag;
[c] Percentage of protein entrapped in resealed erythrocytes;
[d] Mean ± SD for 13 measurements.

molecular weight compounds was more effective than the incorporation of the enzyme (in the latter case the maximal incorporation achieved was only 6.2%). At the same time, in contrast to the low molecular weight compounds, the enzyme was practically not released from erythrocytes at sotrage. In all cases the methods of hypotonic treatment were more effective than chlorpromazine-induced endocytosis.

A comparison of various methods of immobilization of drugs and enzymes in red blood cell "ghosts" has been carried out [385], comparing three methods for harvesting human erythrocyte ghosts. The authors obtained the ghosts under hypoosmotic conditions by direct dilution of cell suspensions and dialysis, and also under isoosmotic conditions by reversible electrical perforation. Red blood cell ghosts, obtained by dialysis, were closest to intact cells. This method also permits the incorporation of enzymes like urease (molecular weight 492 000 D) into erythrocytes with efficiency of more than 10%. Decrease in the molecular weight of an enzyme leads to increase in the incorporation efficiency and for amylase with a m.w. of 45 200 it reaches 65%. Ghosts obtained by dialysis possessed also the highest stability in mouse and dog blood—in dogs the half-life of ghosts was up to 7 days, compared to 18 days for normal cells.

The comparison of the carrier properties of erythrocytes obtained from various animal species is of interest. A comparative study of the ability of different erythrocytes to incapsulate the surrounding medium when the dialysis method of loading is used has demonstrated that this parameter varies considerably—from 20% for rabbit erythrocytes to 70% for dog arythrocytes (see Table 13). A comparison of canine and bovine erythrocytes by the efficiency of incapsulation and by the subsequent pharmacokinetics in vivo has been reported [379]. In [372], the authors have shown the possibility to use nucleated chicken erythrocytes for the immobilization of β-galactose by the hypotonic loading

Table 13. Encapsulation in erythrocytes of several mammalian species by hypotonic dialysis. (From DeLoach J.R., Biblthca haemat, 51, 1, 1985, with permission)

Species	Maximum % encapsulation	Average % encapsulation	Average % cell recovery
Dog	70	40–45	80–90
Mouse	45	30–35	60–70
Human	60	30–40	60–70
Cow	55	35–40	50–65
Pig	55	30–40	50–65
Horse	40	30–40	50–65
Rabbit	20	8–15	30–40[a]

[a] Rabbit erythrocytes often gel during the annealing phase of encapsulation.

method and the similarity in the properties of the preparations obtained and analogous preparations with mammalian erythrocytes.

Evidently, the most promising application is by immobilization of autologous human erythrocytes with subsequent administration of the preparation to the same patient. As for the optimal method of loading, most authors prefer the gentle method of hypoosmotic dialysis.

The largest number of studies with erythrocyte ghosts are devoted either to cellular reactors aimed at the conversion of certain substrates in the blood without leakage of the enzyme and its entrance in body cells, or, to obtain red blood cell-based enzymatic preparations for the intracellular delivery of the corresponding enzymes to the liver for the correction of certain metabolic disorders.

Asparaginase is the most typical representative of the first group, and there are numerous studies on its immobilization in red blood cell ghosts. We will mention only some of these to give a general idea on the state of art. One of the first attempts to incorporate asparaginase in erythrocytes in order to reduce allergic reactions, to prolong the life-span of enzyme activity in the blood flow and to target the enzyme to cells of the reticulo-endothelial system has been described [37]. Using ^{51}Cr-labeled erythrocyte preparations and ^{125}I-labeled asparaginase, the authors have demonstrated that immobilized asparaginase is capable to function in the blood flow of experimental monkeys for 19 and more days, compared to 10 days for the free enzyme (see Fig. 30). At the same time, the authors have found some decrease in the life span of the loaded erythrocytes compared with intact ones (see Fig. 31), and also a decrease in the internal volume of erythrocytes at loading by 30%. Red blood cell ghosts, containing immobilized asparaginase, have been used for experimental therapy of lymphoma 6C3HED in mice C3H [396]. The authors have demonstrated that after the administration of the enzyme in erythrocytes into the circulation of mice, asparaginase activity in plasma could be observed for two weeks, resulting in the complete disappearance

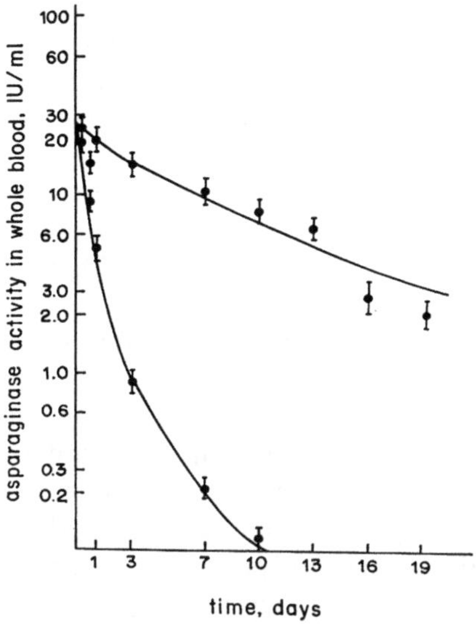

Fig. 30. Asparaginase concentration in blood. (●) asparaginase in RBC's, (▲) native asparaginase. Each point represents the asparaginase activity in whole blood (plasma plus RBC's) drawn from three monkeys (Reproduced with permission from Updike SJ et al. (1976) Science 193:681)

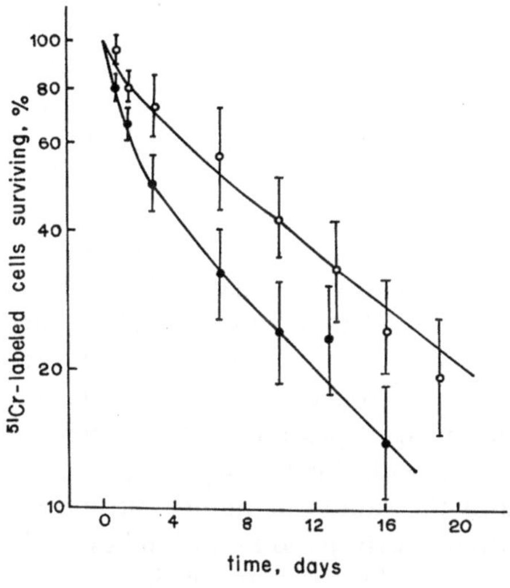

Fig. 31. Survival of enzyme-loaded and control ^{51}Cr-labeled red blood cells (RBC). (○)—data for control unlysed RBC's that were washed and incubated identically as the released RBC ghosts; (●)—data for enzyme (asparaginase) loaded RBC's with asparaginase concentration and activity of 3.9 mg and 780 I.U., respectively, per ml of RBC ghost volume. Each point represents blood samples drawn from three monkeys (Reproduced with permission from Updike SJ et al. (1976) Science 193:681)

of asparagine from plasma. Besides, the authors have found that immobilized asparaginase is able to decrease the level of glutamine both in vivo and in vitro.

In a detailed study of the properties and possibilities of clinical application of asparaginase, immobilized in erythrocytes, it has been shown that in monkeys, autologous erythrocytes loaded with asparaginase and returned to the blood flow induce a stable and prolonged decrease of asparagine content in the blood (2 times longer than with the native enzyme) [391]. The enzyme was used at a dose of 1850 U/kg body weight. In the same study, it has been demonstrated that the immobilization in red blood cell ghosts sharply reduces immunological complications, accompanying the use of enzyme preparations. It was found that when immobilized asparaginase was used, anaphylactic phenomena occur only at levels which were 10-fold higher than those of the native enzyme causing the same complications. The authors believe that the preparation is quite suitable for the treatment of patients with acute lymphoblastic leukemia.

The attempt to immobilize the reactive derivatives of asparaginase—*p*-benzoquinone-asparaginase on the erythrocyte surface is an interesting approach in the field of immobilization of asparaginase on erythrocytes [297]. The activity of the enzyme is preserved at the level of 50% of the baseline. There are no noticeable changes in the properties of erythrocytic membranes after their treatment with modified asparaginase.

Another large field for application of enzymes, immobilized in or on erythrocytes, can be the therapy of diseases associated with hereditary disorders of lysosomal enzymes in liver cells (see 4–6). This trend is under development for more than 15 years, and a study [368] attempting to immobilize β-glucosidase and β-galactosidase in erythrocyte ghosts by the rapid hemolysis method is one of the first. The clinical trial with glucocerebrosidase immobilized in erythrocyte ghosts on a volunteer with Gaucher's disease was described [398]. The enzyme was purified from human placenta with a 60% yield at 6 000-fold purification. The incorporation of glucocerebrosidase into red blood cell ghosts was carried out by hypotonic dialysis using autologous erythrocytes from the patient. The intravenous application of immobilized glucocerebrosidase did not cause any clinically manifested side effects. A later, detailed study of the therapeutic effect of the erythrocytic preparation [399] has shown that its application induced an improvement of liver function, a decrease in size, and relief of the subjective complaints of the patient. The possibility and the prospects of use of the enzyme immobilized in red blood cell ghosts for the therapy of congenital metabolic disturbances are also indicated by the authors of [400], who outline that primary hyperoxaluria and phenylketonuria are among the potential loads of such therapy. It is also important that the use of erythrocytic preparations sharply reduces immunological complications and makes it possible to administer the preparations repeatedly, as it has been shown with β-glucuronidase [401].

The latter results were also supported by the data from immune titration of native and immobilized L-asparaginase with the corresponding antibodies [402], where a sharp decrease in the inhibition of the erythrocytic preparation from the action of antibodies was shown.

A number of other examples of the possible use therapeutic enzymes immobilized in red blood cell ghosts for the treatment of congenital "errors" of metabolism have been reviewed [392], including the application of erythrocytic urate oxidase, phenylalanine ammonia lyase and glucosidase for the normalization of urea, phenylalanine and glycogen levels in the body (mainly the liver). In the case of phenylalanine ammonia lyase, the efficiency of incorporation is ~13% and the immobilized enzyme does not change its properties compared to the native enzyme and is not released from cells during 2 weeks of storage. In vivo, after intraperitoneal injection of free phenylalanine into mice, the erthrocytic preparation of the enzyme induced a noticeable decrease in phenylalanine levels in the blood, surpassing the effect the native enzyme injected. Normalization in the level of phenylalanine in blood was achieved for less than 2 h (see Fig. 32).

In case of 1,4-β-glucosidase, the efficiency of its incorporation into erythrocyte ghosts was 48%, and the incorporation also caused only minor alteration of the properties of the enzyme. At the same time, erythrocytes with immobilized enzymes induced a more prolonged decrease in the content of glycogen in the liver than the native enzyme (Fig. 33).

The immobilization of phenylalanine ammonialyase in red blood cell ghosts is also mentioned in [385].

The incorporation of other therapeutically significant enzymes into erythrocyte ghosts also has been described. It has been suggested to immobilize uricase from pig liver in erythrocytes for the degradation of uric acid in vivo [403]. For the treatment of hereditary hyperargininemia, it has been suggested to use arginase-loaded erythrocytes [404]. In vitro, incorporation of arginase from bovine liver into human erythrocytes can elevate the rate of arginine utilization in a system containing arginase-free erythrocytes from patients with hyperargininemia.

The immobilization of thiosulphate sulfotransferase in erythrocytes is also of some interest [293]. The essence of this procedure is an attempt to enhance the antidote action of sodium thiosulphate after intoxication with cyanides. It was

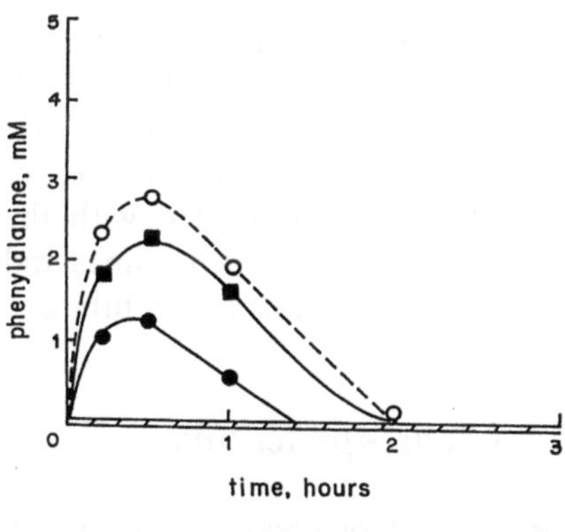

Fig. 32. Concentrations of plasma phenylalanine in mice following intraperitoneal injection of phenylalanine. Additional injections contained 0.9% NaCl-solution (○) and phenylalanine ammonia lyase both as free (■) and encapsulated (●) enzyme. The hatched area at the bottom represents the range of normal phenylalanine levels in mice (Reproduced with permission from Sprandel U (1985) Biblthca haemat 57:7 Karger, Basel)

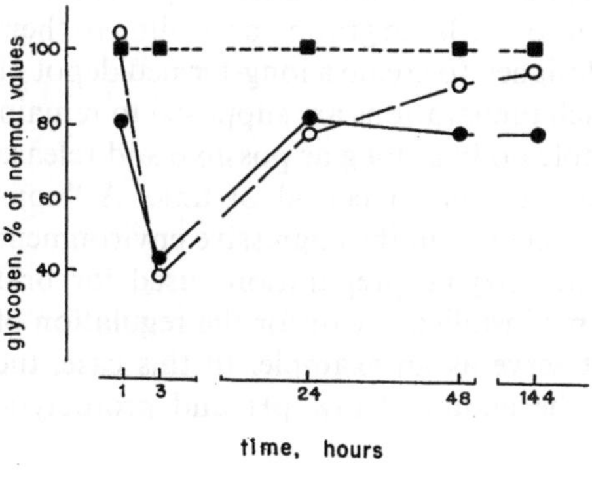

Fig. 33. Total liver glycogen of mice following the injection of both free (○) and encapsulated (●) amyloglucosidase (25 mU per animal). Each data point represents an average of at least six animals and is given as percentage of the normal liver glycogen of mice (Reproduced with permission from Sprandel U (1985) Biblthca haemat 57:7 Karger, Basel)

also suggested to treat glucoso-6-phosphate dehydrogenase deficiency, using either direct incorporation of the enzyme into abnormal erythrocytes [405, 406] or its immobilization in erythrocyte ghosts and subsequent administration of the preparation to the patients.

Attempts exist to immobilize enzymes by incorporation into other cells, in particular, into hepatocytes and leukocytes. In the latter case, it has been reported that these cells can penetrate the blood-brain barrier and deliver drugs to various areas of the brain, which may have a very important clinical significance [407]. Incorporation of therapeutic enzymes into macrophages can also be promising, since macrophages possess the ability to accumulate in areas of inflammation and therefore can transport entrapped enzymes there.

As mentioned above, some studies are devoted to the immobilization of therapeutic enzymes not to the inside of cells, but on their surface. Besides the above experiment with asparaginase [397], glucosidase, arginase and peroxidase have been immobilized on the surface of platelets and erythrocytes by adsorption [408]. Enzymes whch can tightly bind to cells acquired a noticeably elevated stability against thermoinactivation.

Furthermore, it is interesting to mention the attempts to apply erythrocyte ghosts for the delivery of different proteins to other cells, e.g., to cultured rat hepatocytes [409] or hemopoietic human cells [410]. Horseradish peroxidase has been microinjected into mammalian cells in tissue culture by a erythrocyte ghost-mediated technique [410a].

In spite of the significant achievements in the field of immobilization of therapeutic enzymes into erythrocyte ghosts, there are certain difficulties which slow the transfer of methods developed in the laboratory into clinical use. According to the opinion of the authors of [411], the major drawbacks of erythrocytes as carriers of enzymes and other drugs are associated with the difficulties to change their structure, the short life-span of their membranes and their permeability for low molecular weight compounds, toxins and inhibitors.

4.2 Insoluble Immobilized Enzymes for Local Application

Insoluble derivatives of therapeutic enzymes are also of great importance in practical medicine. They may be divided into two large groups, according to their field of application. First, preparations designed to create a long-termed depot of therapeutic enzymes in the organism. Such preparations are supposed to remain at the site of administration or in the whole body as long as possible and release entrapped enzymes at a given rate during a given period of time. A "non-working" part of an enzyme must be protected from the aggressive environment of the organism by a polymeric carrier. Enzyme preparations used for oral administration in the treatment of enzyme insufficiency or for the regulation of enzyme activity in the alimentary tract serve as an example. In this case, the enzyme must preserve stability under the action of low pH and proteolytic enzymes in the stomach.

Second, there is a great obstacle closely related with the problem of extracorporeal therapy, that is obtaining biocompatible materials for prosthetics of internal parts of the body. One aspect relates to vascular prostheses which, without immobilization of thrombolytic enzymes on their surface, cause increased blood coagulation. The same is true for catheters designed for prolonged stay in vessels.

Enzymes, immobilized on the solid implant surface (prosthesis), should not be released into the environment to express their activity, but should remain on the surface of the material for a required period of time. These materials (mainly polymers) for prosthetics are well established and the main problem lies in the development of optimal and effective methods of therapeutic enzyme immobilization on their surface.

4.2.1 Slow Release Systems and Their Classification

Systems of controlled drug release will be considered here according to the classification suggested by Bangall [412]. First, let us define the controlled release of a drug as "delivery of a certain portion of the drug at a given rate into an required organ". It should be pointed out that the term "a required organ" includes also the blood flow and the body as a whole, if the active preparation is supposed to be delivered into various tissues of the body from a certain place.

The principal types of controlled drug release systems are presented in Table 14. In reality, the picture is more complicated: in each case both the kinetics (or profile) of immobilized drug release from the carrier and its possible localization in a required area should be taken into consideration. The following systems of slow drug release have been recognized [412]: degradable systems, when an entrapped agent is released as a result of the destruction of matrix material in the organism (in this case the geometry of a particle or an implant is of particular importance); non-degradable systems that can release the drug (enzymes as well) either in a passive way due to diffusion determined by the difference in the drug concentration inside and outside the device, or in an active way due to some outer force (of special interest is the case when molecules of the therapeutic preparation are incorporated into non-permeable polymeric gel

Table 14. The principal types of controlled drug delivery systems

1. Minipumps	(a) mechanical (b) osmotic
2. Polymeric biodegradable matrix (drug release is determined by chemical processes of polymer degradation)	(a) drug is mechanically encapsulated in polymer (b) drug is chemically bound
3. Microreservoirs (drug release is determined by diffusion via a capsule membrane or container destruction)	(a) microcapsules (polymers, liposomes, cells) (b) nanoparticles

together with ferromagnetic particles, and after gel implantation a high-frequency magnetic field is applied to the gel localization which provokes vibration of magnetic particles, changing the parameters of gel cells and providing free release of the entrapped preparation [413]; various devices and materials of intermediate or mixed type are also possible. Here we shall not dwell on possible mechanisms of carrier destruction or passive release of entrapped compounds. These mechanisms are numerous, including usual proteolytic splitting of carriers or swelling of polymeric material when the preparation "leaks" as a result of gradual enlargement of pore size.

4.2.2 Polymeric Enzyme-Containing Implants and Microparticles with Immobilized Enzymes

Microparticles of ion-exchange resins proved to be promising carriers for obtaining stabilized enzyme preparations for the treatment of alimentary tract diseases. Thus, protein preparations can be stabilized by cross-linked polymeric electrolytes via polyfunctional coupling into the structure of insoluble enzyme/polyelectrolyte complexes. α-Amylase binding with cross-linking polymers containing methylacrylic acid residues results in a drug, resistant to acid and gastric juice action [414]. In some cases, the use of appropriate ion-exchangers produces preparations capable of releasing bound enzymes at a definite pH value. In particular, macroporous ionites are used for reversible protein binding. An increase in pH can bring about a complete destruction of the macroporous ionite-insulin insoluble complex, with active insulin being released [415].

Enzymes which catalyze the conversion of low molecular weight substrates can be used in the form of an enzyme-containing polymeric gel. Such a gel will protect the enzyme from the aggressive biological medium, and will not hinder diffusion of substrates to the enzyme and the release of reaction products. Asparaginase, immobilized in the microparticles of polyacrylamide gel, has been introduced parenterally (as a saline suspension) into experimental mice with asparaginase-sensitive lymphoma. This resulted in the decline of asparagine concentration in the blood from 75 to 1.5 nmol/ml, although 8 days later it returned to normal levels [416]. Similar results were obtained studying asparaginase entrapped into gel microparticles made from poly-(2-oxylmethyl)-metacrylate. Uratoxydase incorporation into polyacrylamide gel, hydroxymethylmetacrylate gel or into gel formed by polyamine and polyisocyanate produces an enzyme preparation with high activity and increased stability [417, 418].

It has been already mentioned that enzyme immobilization on insoluble polymeric carriers is very convenient for the development of preparations that can be deposited at the site of injury for a sufficiently long period of time, with continuous release of the active drug into the environment to achieve a high local concentration. An insignificant total dose of an enzyme permits to achieve its high local concentration.

Such drugs can be deposited in a defined part of the body or via vessels using catheters, or by subcutaneous/intramuscular implantation (see a schematic picture on Figs. 34 and 35).

There are several approaches to obtain immobilized enzymes with controlled release of active material for parenteral administration.

According to one of the trends developed [419–422], biologically active macromolecules (lysozyme, catalase, trypsin protein inhibitor, etc.) are introduced into polymeric tablets or granules obtained from various synthetic biocompatible polymers: polyacrylamide, polyvinylpyrrolidone, polyvinyl alcohol, copolymers of ethylene and vinylacetate etc. At first, a polymer solution at an appropriate concentration is prepared in water or in a mixture of water and non-denaturating organic solvent. Then, the appropriate amount of a protein is added to the polymer solution. The mixture obtained is placed into a vial and the final product is obtained by vacuum evaporation of the solvent. Cross-linking reagents can be introduced into the system if needed. By varying concentrations

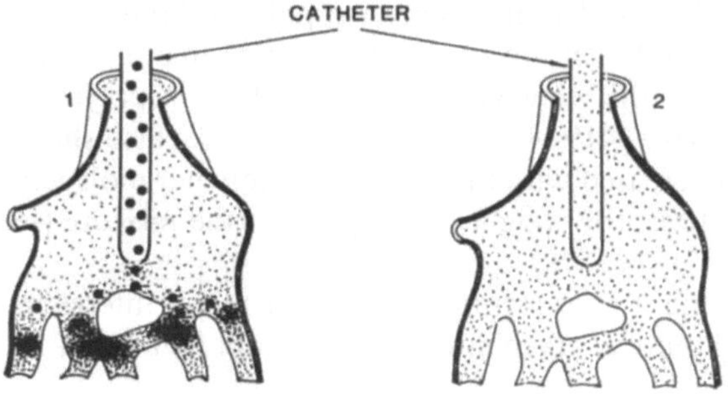

Fig. 34. Method of obtaining a drug depot in the organism by intravascular administration of enzyme-containing microparticles via a catheter (*1*—the immobilized enzyme permits to create a high local concentration, *2*—the native enzyme is distributed all over the circulation) (Reproduced with permission from Torchilin VP (1987) Adv Drug Deliv Rev 1:41)

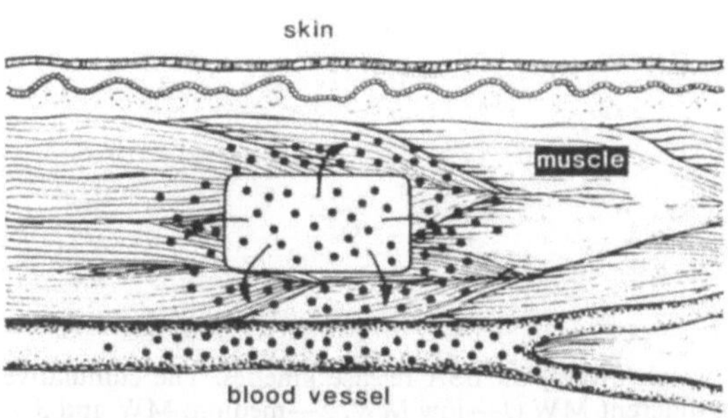

Fig. 35. Method of obtaining a drug depot by intramuscular implantation of enzyme-containing polymeric material. The enzyme is released from the pellet and creates a high local concentration; it can also reach the circulation (Reproduced with permission from Torchilin VP (1987) Adv Drug Deliv Rev 1:41)

and polymer-to-protein ratio, one may obtain granules which are able to release an active substance into the environment at a practically constant rate for time intervals starting at several hours to several months. The enzyme is completely protected and retains its activity as long as it stays inside the granule.

The most impressive result of was obtained when polymeric granulae containing the tumor angiogenesis factor were used. Implantation of such granulae into the rabbit cornea induced vascularization of adjacent sections of the cornea. Granules preserved their activity for at least 40 days. Under natural conditions, this factor induces proliferation of vessels towards the tumor to provide it with nutrients in the blood. Newly formed vessels remain in the tissue for some time after the removal of the granulae with the immobilized growth factor from the cornea; they gradually disappear by the end of the third week. Vascularization does not occur if the inhibitor of the growth factor was immobilized into other granules and simultaneously implanted near by.

The ability of polymers used to induce an inflammation at the site of implantation has been studied in detail. This is important, since a "theoretical" polymer and a real product may have different toxicity, e.g. polyacrylamide is not toxic and causes no inflammation, but it always contains some free monomer (acrylamide) which is toxic. From this viewpoint, the evaluation of polymeric carrier biocompatibility is necessary [419–422]. These authors have also shown that the rate of protein release decreased with the increase of polymer molecular weight even at equal concentrations (by weight) of polymer in the mixture (see Fig. 36).

The approach developed in the studies mentioned above is rather interesting and promising. Three aspects should be taken into consideration: 1) enzymes are

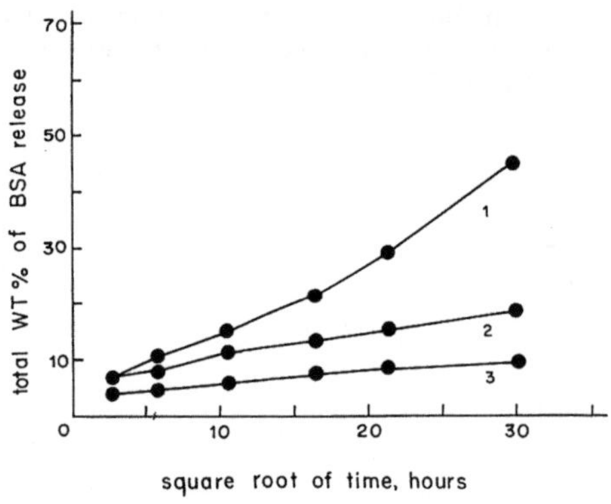

Fig. 36. The effect of polymer molecular weight (MW) on BSA release kinetics. The cumulative percentage release of BSA from systems of different MW (*1*—low MW, *2*—medium MW and *3*—high MW). Carriers at the same polymer solution concentration (10 g/dL). The loading and particle size of BSA are 25% and 106–150 μm, respect. Release kinetics were studied at 37 °C in phosphate buffer solution, pH 7.4. Each point represents the mean of five samples. Standart deviations were < 10% of the respective values. (Reproduced with permission from Hsu TT-P, Langer R (1985) J Biomed Mater Res 19:445. Copyright 1985, John Wiley and Sons)

released from the polymer in their native form and can undergo fast inactivation; 2) slow release of the native protein can enhance the immune response of the organism, since in this case the carrier may be considered as an immunological adjuvant; 3) finally, the fate of the polymeric carrier remains unclear after complete release of the active material.

Carriers may be used that can be destroyed under physiological conditions yielding products easily released from the body. If the rate of carrier degradation is close to the rate of enzyme release, the conjugate as a whole disappear simultaneously with completion of the therapy course. In this case, the enzyme can be covalently bound to a solubilized carrier fragment that will stabilize an enzyme in the solution and decrease side effects caused by the released compound. Enzyme preparations, immobilized on various polysaccharides, have been proposed for the creation of such drug depots [45, 424–426]. As already indicated, partial oxidation of polysaccharides by periodate groups gives reactive derivates which easily interact with protein amino groups. Naturally, such a method is suitable for both soluble polysaccharides and their insoluble derivatives, e.g., Sephadex and Sepharose, which are cross-linked polysaccharides. The oxidative activation of cross-linked dextran (Sephadex) has another important effect: after oxidative modification, Sephadex acquired the ability for slow dissolution in water. The rate of the dissoclation may vary from hours to weeks and is regulated by the degree of Sephadex oxidation [424, 425]. Various therapeutic enzymes have been immobilized on such carriers, such as thrombolytic enzymes (fibrinolysin, urokinase and streptokinase). It is possible to bind 10–80 mg of active enzyme per g of carrier depending on the degree of carrier activation, enzyme:carrier ratio and duration of immobilization reaction. After complete solubilization (together with the stabilizing carrier fragment), the enzymes revealed unchanged activity and increased stability. Animal experiments demonstrated the low antigenicity of such preparations. It was shown that thrombolytic enzyme preparations, obtained by this method, can be highly effective for local treatment of thromboses and thromboemboli in any vessel which can be reached by catheterization. Thus, the suspension of fibrinolysin immobilized on partially oxidized Sephadex locally deposited at the site of a thrombus in experimental thrombosis of a dog femoral artery [428, 429] caused complete lysis of the thrombus and restoration of the blood flow within one after injection (Fig. 37). At the same time, the thrombus was still present in the second (control) artery, indicating that the fibrinolytic activity of the systemic blood flow remained unchanged. Lysis was achieved when the dose of the active enzyme used was $\sim 1\%$ of the amount of fibrinolysin necessary for lysis of the analogous thrombus after routine intravenous administration of the native enzyme. The carriers developed can retain immobilized proteins even in the heart muscle for 24 h and longer without any harmful consequences for the myocardium [429, 430].

If immobilization conditions are chosen correctly, the kinetic properties of thrombolytic enzymes change insignificantly when transferred into an insoluble state [431, 432].

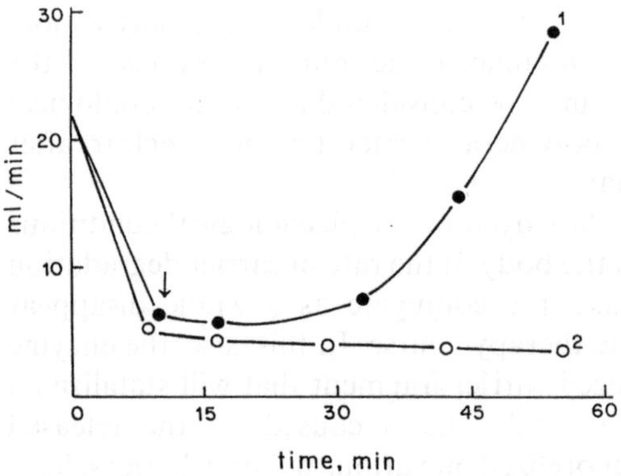

Fig. 37. Local action of immobilized plasmin in dogs with experimental thrombosis of femoral arteries (Reproduced with permission from Chazov EI et al. (1978) Thromb Res 12:809). Microparticles with 1000 IU immobilized plasmin were administered via catheter directly to the thrombus in one artery. Blood flow in treated (*1*) and untreated (*2*) arteries. *Arrow* indicates onset of complete occlusion and immobilized enzyme administration

One of the drawbacks of the approach described is the impossibility of obtaining polysaccharide-based preparations which can survive in the body more than several days and the necessity of preliminary activation of the carrier.

Other studies of heterogeneous preparations of immobilized enzymes for medical application are devoted to the development of microspheric reactive carriers, based on synthetic polymers of the polyacrylamide type or acryloyl polysaccharide derivatives, e.g., acryloyldextran [433–436]. The authors obtained microspheres (size 0.25–0.30 μm) from highly cross-linked polyacrylamide by emulsion polymerization and investigated their biodistribution in mice and rats after intravenous and intraperitoneal injections. It was demonstrated that a major amount of the preparation after intravenous injection is accumulating in the liver and spleen. The radioactive label, associated with a polymer, gradually disappeared from the body within 4 months. It should be pointed out that the carrier was capable of slow biodegradation and no toxic effects were observed during 45 weeks after polyacrylamide microspheres administration at a total dose of 9.8 mg of polyacrylamide in mice [433]. Local impairment by microgranules noted in the mouse liver and in macrophage cultures disappeared within 2–3 weeks, without any complications [434].

Polymerized acryloylstarch microparticles are able to trap up to 40% by weight of a protein [435]. The authors demonstrated that these microparticles can be totally degraded by the action of amyloglucosidase and may be considered as promising carriers [436]. The authors also mentioned the possibility of enzyme introduction into a complex solution, containing both a synthetic polymer of the polyacrylamide type and a modified polysaccharide, with the subsequent preparation of an enzyme containing microparticles from the above mixture. The gradual degradation of the polysaccharide component of the polymeric gel and slow washing out of the polymer from the "releasing" polymeric mixture is supposed to occur under physiological conditions.

In vivo experiments with the administration of polyacrylamide microspheres containing immobilized asparaginase into mice demonstrated that the immobilized enzyme was able to effectively decrease asparagine levels in the blood of experimental animals [437]. An intravenous injection of the preparation (at doses of 200 U/kg) lead to the decrease of plasma asparagine below 10 nmol/ml, although asparagine levels is normalized within 4–5 days in the case of administration of both immobilized and native enzymes. At the same time, the intraperitoneal injection of such microparticles (at an enzyme dose of 1.000 U/kg) induced a prolonged decrease in asparagine levels in the blood flow for 2 weeks. The intraperitoneal injection of the native enzyme yielded no effect at all. The therapeutic effect of the enzyme immobilized on polyacrylamide microparticles against 6C3HED lymphoma in mice has also been shown [416], where the enzyme was able to exert a therapeutic effect without leaving the particle. The optimal therapeutic effect (total tumor regression) was observed in these experiments after intraperitoneal or intramuscular injection of immobilized asparaginase at a dose of 2 to 20 U per kg body weight.

Similar results were obtained using enzymes immobilized on microspheres made of biodegradable polyacryl dextran [439, 440]. Immobilized dextranase and carbonic acid anhydrase dramatically increased their thermostability without any changes in catalytic properties (unchanged K_M and V_{max} values). In vivo experiments demonstrated that dextranase activity accumulated in liver and spleen lysosomes. The authors also succeeded in demonstrating the accumulation of dextranase, immobilized on microparticles on dextran, in lysosomes. The native enzyme proved to be ineffective. The authors showed specifically that the immobilized enzyme really penetrated cell lysosomes with pathological dextran accumulation and there exerted its effect [441].

The authors of [442] have used epoxy-activated polyacrylate granules for the immobilization of β-galactosidase on their surface. Preparations were obtained containing 40 mg of a protein per g carrier; the immobilization yield was 42%. β-D-galactosidase immobilized on agarose cyclic imidocarbonate under in vitro conditions is capable to convert glycosphyngolipid—the principal substrate accumulated during Fabri's disease; unfortunately, the authors failed to observe an effect in experiments in vivo [443]. Cross-linked polyvinyl alcohol also may serve as carrier for the immobilization of therapeutic enzymes. It was already successfully used for the immobilization of some low molecular weight cytostatic agents [444]. This is also true for polyalkylcyanoacrylate [445].

Cellulose is a natural polymer capable for biodegradation in the body. Its various derivatives [446], as well as products of partial oxidation [447], have been used for the immobilization of model drugs such as α-chymotrypsin. Enzyme immobilization on insoluble microparticles obtained from compounds typical for the organism itself appear to be a relatively novel approach. For instance, lipase and amilosubtiline have been immobilized on insolubilized globulins [448]. It is interesting to note that in this particular case the immobilized enzyme can be transferred from the pellet into solution and vice versa simply by altering the ionic strength of the solution.

The product of leukocytic elastase inhibitor immobilization on albumin microspheres has been proposed for the treatment of emphysema [449]. In this case, an interesting method for inhibitor immobilization via spacer arms of various length was used. The optimal activity of the inhibitor was achieved when a spacer with the length of more than 24.3 Å was used. Each albumin molecule can bind up to 30 molecules of the inhibitor. Microspheres obtained from this modified albumin in experiments in vivo accumulated well in the rat lungs, their half-life being about 17 days.

Lactoperoxidase, exhibiting good anti-microbial properties, was covalently immobilized on BrCN-activated Sepharose 4B and on concanavalin A-Sepharose 4B [450]. There are data available on the use of solid carriers for immobilization of various multienzyme systems which retain their activity [451].

To avoid the loss of enzyme activity or distortion of their normal catalytic properties, some authors suggested the use of enzyme immobilization in a structure of polymeric gels which permit the maximally retention of the native conformation of a protein globule [452–456]. The principle of this method is to obtain the copolymerization of an enzyme with a synthetic monomer. As far as the proteins contain no groups capable to enter the copolymerization reaction, such groups can be artificially introduced into the enzyme molecules by modification of protein reactive groups with derivatives of unsaturated and polymerizable compounds (acryloylchloride is one of the most frequently used modificators). The derivates obtained, usually called "macromonomers", successfully enter into copolymerization reactions with traditional monomers and when additional cross-linking agents are introduced into the system, the derivates form an insoluble polymeric material exhibiting enzymatic activity.

Some of the recent papers devoted to the problem of insolubilized therapeutic enzymes are of special interest. Prostaglandin-N-synthetase from the microsomal fraction of sheep vesicular glands was non-covalently entrapped into polyelectrolyte complexes based on such polyions as poly (C), chondroitinsulfate, polylysine, and gum-arabic [457]. The authors assume that these preparations can be used in practical medicine to control prostaglandin metabolism in the body.

Immobilized urokinase preparations in the form of tablets, capsules and suppositories have been tested in rats and dogs [458]. The authors assume that the composition suggested, containing fatty acids and polyalkylenglycols as enzyme stabilizers, can be used for the development of original forms of thrombolytic drugs for oral application.

From the theoretical point of view, some recent studies (e.g., [459]) appear to be of importance, concerning the role of the polymeric matrix structure and the interactions between separate particles in drug release from polymers by diffusion.

4.2.3 Enzymes for Modification of Polymeric Implants and Prosthetic Devices

As already mentioned, various polymeric implants may be combined with insoluble preparations of immobilized enzymes, such as vessel prostheses, the surface of which has been modified with an appropriate enzyme to provide maximal biocompatibility (decrease of thrombogenicity). Various thromboresistant materials may improve the possibilities of artificial organ implantation (cardiac valves, blood vessels, artificial pancreas, etc.). Polymers with enzymes immobilized on their surface (e.g. enzymes preventing platelet adhesion and activation of fibrinolytic enzymes) have proved to be the most promising materials.

Covalent immobilization of proteases, capable of catalyzing the hydrolytic degradation of fibrin (trypsin, chymotrypsin, fibrinolysin, urokinase and others) on the surface of various polymeric materials, leads to a sharp decrease in thrombus formation. Thus, surfaces of artificial polymeric blood vessels made of polycarbonate, polymethylmetacrylate and polyvinylchloride become athrombogenic when covered with urokinase immobilized by adsorbtion, with subsequent covalent coupling via glutaric dialdehyde [460]. The same is true for nylon surfaces containing urokinase immobilized via polyethyleneimine or copolymers of maleic anhydryde and methylvinyl esters [461]. The clinical efficacy of nylon tubes modified with urokinase was assessed in 32 patients who underwent surgical intervention. It was found that the tubes used in vivo retained the ability to activate plasminogen for 23 days and could be used repeatedly [462]. The authors recommend the use of nylon tubes with immobilized urokinase to build parts of artificial organs, artificial vessels, intravenous catheters and surgical drainages.

Vessel prostheses can be also used as carriers for the immobilization of many other physiologically active substances. Thus, a dacron prosthesis with immobilized asparaginase produced a drastic decrease in asparagine levels in the blood flow with a simultaneous decrease in antigenicity of the immobilized enzyme [463].

The implantation of collagen prostheses made of bovine carotid arteries and containing immobilized asparaginase, into various regions of the thoracic and abdominal aortam the dog lead to prolonged decrease in asparagine content in the blood plasma (for up to 6, 5 days) [464]. Determination of asparaginase activity in prostheses removed from the animal after 6–15 days after implantation demonstrated a decrease in asparaginase activity by 7 to 70%, depending on the method of enzyme immobilization.

Interesting fibrinolytically active polymers and methods for their preparation have been described [465]. Polymers which are usually used for prosthesis manufacturing have been modified with plasmin, papain, trypsin, streptokinase and chymotrypsin. Enzymes have been immobilized on polyamide prostheses via glutaric dialdehyde after preliminary activation of polyamide by partial hydrolysis with HCl, leading to the appearance of free amino groups on the polymer surface. Streptokinase or urokinase have been immobilized on the surface of silicon rubber via cyanogen-bromide activation of free hydroxy groups. The authors have also produced polyethylene catheters modified with thrombolytic enzymes and investigated the stability of fibrinolytic activity in catheters with immobilized plasmin. The activity of polymer-immobilized enzyme was determined by the size of lysis zones on fibrin-plasminogen-agarose gel. It was demonstrated that the immobilized fibrinolytic activity is stable enough both in vitro and in contact with the human blood serum.

Although successful results have been achieved in the studies mentioned, these do not highlight the full potential of this approach due to the fact that the traditional methods of immobilization do not provide enzyme fixation on the polymeric surface in the native conformation and often induce a pronounced decrease or changes in immobilized enzyme activity. The copolymerization of enzyme macromonomers with the appropriate "usual" monomer directly on the surface of a polymer may help to solve this problem. The formation of an enzyme-containing layer on the surface of the polymeric implant has one more advantage: this layer may be a hydrogel, i.e. a "macroporous" polymeric network, containing a great amount of water. The presence of such a layer on the prosthesis surface makes the contact of blood with the foreign material less affecting and more physiological.

The methods of preparation and isolation, and the properties and use of macromonomers of fibrinolytic enzymes to obtain athrombogenic polymeric materials for prosthetic devices have been described in detail [466]. Here we shall dwell on the most important results of this study, which involved both specific enzymes of the blood fibrinolytic system and enzymes exhibiting general proteolytic activity.

Copolymerization of trypsin and plasmin macromonomers with acrylamide and N,N′-methylenebisacrylamide yields hydrogels with covalently immobilized trypsin or plasmin, in which enzymes retain their catalytic activity and are capable to hydrolyse different proteins in solution, including blood plasma proteins. Protein macromonomers have been obtained by enzyme modification with acrylic acid chloride. Table 15 shows that the modification has practically

Table 15. Activities of native physiologically active compounds (I) and of compounds acylated with acryloyl chloride (II) [466]

Physiologically active compound	Activity: Substrate	Type of activity, units	I	II
Trypsin	N-benzoyl-L-arginine ethyl ester	esterolytic activity, ueqv. substrate/mg/min	41 ± 2	45 ± 2
Fibrinolysin	—"—"—"—	—"—"—"—	34 ± 2	31 ± 2
Trypsin	casein	proteolytic activity, umol tyrosine/mg/min	2.38 ± 0.05	4.92 ± 0.07
Trypsin inhibitor from soya beans	trypsin	antitryptic activity IU/mg[a]	1.16 ± 0.04	1.11 ± 0.04
Serum albumin	organic dyes	binding activity, mol^{-1}[b]	$(0.40 \pm 0.10) \times 10^8$	$(0.32 \pm 0.17) \times 10^8$

[a] IU—inhibitory unit, the inhibitor quantity inactivating 1 mg of trypsin.
[b] —binding constant for organic dyes.

Table 16. The dependence of enzymatic activity of modified polypropylene on the composition of grafting mixture (irradiation dose 0.3 Mrad, substrate—1% casein solution) [466]

Weight ratio acrylamide: unsaturated enzyme in reaction mixture	Activity, ug tyrosine/min/cm²	
	Fibrinolysin	Bacterial proteinase
0	0.05 ± 0.01	0.07 ± 0.01
200	0.47 ± 0.03	0.96 ± 0.03
400	2.00 ± 0.10	2.78 ± 0.12
800	2.16 ± 0.11	2.70 ± 0.12

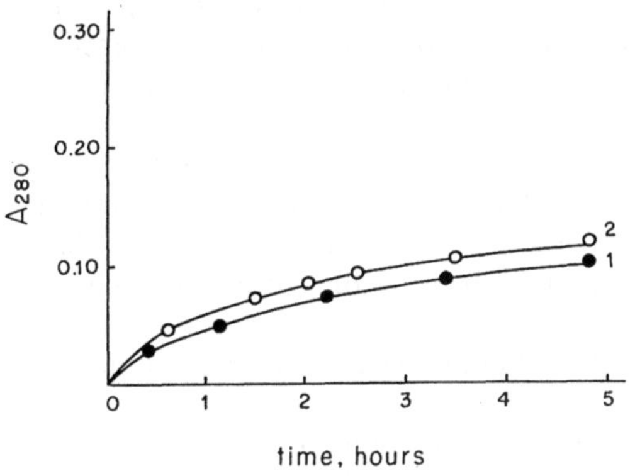

Fig. 38. Hydrolysis of fibrinogen (*1*) and albumin (*2*) by trypsin, immobilized in polyacrylamide gel (0.12 mg trypsin/ml of swollen gel). (Hydrolysis expressed as OD_{280} increase with time) [466]

no effect on protein properties, and the data from Table 16 demonstrates that variations in hydrogel monomer composition permit changes in the amount of active enzyme immobilized on hydrogel-modified surface of biomedical polymers.

Figure 38 demonstrates the results of the study on fibrinogen and serum albumin hydrolysis by trypsin immobilized in polyacrylamide hydrogel. As seen in this Figure, the rate of hydrolysis of both proteins is the same, i.e. the selectivity of immobilized trypsin towards a specific protein is missing. Since the size of the hydrogel matrix pores is large enough (on average, 100–150 nm) for both substrates to diffuse freely to the active site of the enzyme, it has been assumed [467] that the hydrolysis rate of one of the proteins may increase if its concentration in the vicinity of the active site is considerably higher than of another one. To achieve this, a specific ligand was proposed, which when immobilized in the same hydrogel matrix in the vicinity of the enzyme, can bind one of the proteins.

If heparin, an acidic mucopolysaccharide exhibiting a high capacity for complexing with fibrinogen, is added to hydrogel during trypsin immobilization, the rate of hydrolysis increases significantly. Figure 39 shows that the quantity of

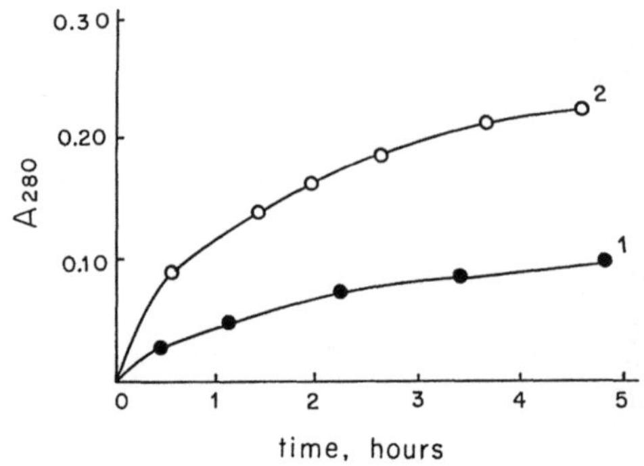

Fig. 39. Hydrolysis of fibrinogen by hydrogels, containing immobilized trypsin (*1*; 0.12 mg/g) and trypsin + heparin (*2*; 0.12 mg/ml trypsin + 2.95 mg/ml heparin). (Hydrolysis expressed as OD_{280} increase with time) [466]

hydrolyte products depends on the duration of contact of the fibrinogen solution with binar hydrogels containing covalently immobilized heparin and trypsin. It is seen that the binar hydrogel degrades protein substrate more actively, compared with hydrogel containing trypsin alone, the reason for this being the increase in local fibrinogen concentration during the reaction.

Consequently, the same immobilized enzyme can predominantly convert defined substrates in the reaction mixture. This is achieved by introduction of a ligand specific to target substrate into the enzyme vicinity which acts like a specific substrate binding site, increasing the substrate concentration in the layer of hydrogel.

For the first time, the potential of an immobilized enzyme (urokinase) to increase polymer hemocompatibility have been well documented in [460]. Polymeric surfaces (polycarbonate, polymethylacrylate or polyvinylchloride) were preliminarily coated with a graphite layer, followed by enzyme adsorption. Urokinase was additionally cross-linked with glutaric dialdehyde to prevent its rapid washing out into the blood flow. Determination of the fibrinolytic activity of immobilized urokinase in vitro (dissolution of coagulated plasma) and in vivo (implantation of modified polymers into animal aorta) demonstrated that all modified polymers exhibited the capacity to dissolve fibrin and prevented thrombus formation on the polymer surface within two months. In control experiments, thrombi on the surface of non-modified polymers were formed on the 5th day of implanation.

The mechanism of increasing polymer hemocompatibility by modification with proteolytic enzymes has rarely been discussed. The reason for this phenomenon always appeared obvious enough: immobilized enzymes directly or via plasminogen activation catalyse the lysis of fibrin clot. Besides, one may assume that immobilized proteinases excert a general effect on the coagulation process and proteolytic enzymes.

A detailed study using two different enzymes, plasmin and proteinase from Streptomyces spheroides, completely supported this assumption [468]. Immobilization of the above enzymes on the polymer surface by radiation grafting of the mixture of a hydrophilic monomer and enzyme macromonomers leads to a significant decrease in coagulability on contact with the modified polymer, even when enzyme release into the blood flow does not occur.

One of the reasons for this phenomenon is the decrease of platelet adhesion on modified polymers and the prevention of degradation of adhered platelets. Thus, if the relative index of platelet adhesion (equal to the ratio of platelet quantity adhering to the surface under investigation and a glass surface) for non-modified polyethylene has the value of 1.8, this value decreases to 1.1 for modified polyethylene. Besides, electron microscopy has demonstrated that adhering platelets degrade far less when compared to adhesion on non-modified polymer surfaces.

In many cases, heparin has been immobilized on polymers to achieve athrombogenicity. However, heparin excerted no effect on such events as platelet adhesion and the lysis of stabilized fibrin. Since enhanced platelet adhesion on the surface of heparin-containing polymers is caused by fibrinogen concentration on this surface, due to its complexing with immobilized heparin, a joint immobilization of heparin and proteases capable to lyse adsorbed fibrinogen was proposed as one of the modes to decrease platelet adhesion on a polymer surface [469].

Currently, soluble complex preparations based on heparin and proteinases are used in clinical medicine for the treatment of thromboses. The injection of fibrinolytic enzymes or their mixtures with heparin into the blood brings about a partial or complete thrombus lysis, due to the plasmin-activating and lytic action of these enzymes. The data obtained show that heparin alone exhibits no lytic activity and in a complex with streptokinase lyses the thrombus only partially [470].

Improved hemocompatibility (as compared with heparin-containing polymers) of polymers modified by both heparin and trypsin has also been observed in vivo and in vitro [466]. The dynamics of accumulation of thrombotic masses on the surface of polyethylene catheters attached by femoral arterial-venous shunts in dogs has been assessed. The quantity of thrombotic masses on the surface of non-modified polyethylene amounted to 8–10 ug/cm^2, while this value decreased to 1–3 ug/cm on the surface of polyethylene modified by heparin, trypsin or their mixture (exposition time 9 min). Differences in blood compatibility of the materials used are also quite evident when the contact of a catheter with blood is prolonged to 20 min. Thrombotic masses on the surface of polyethylene and polyethylene modified by heparin or trypsin separately amounted to 10–14 ug/cm^2, and on the surface of polyethylene modified by trypsin/heparin mixture amounted to 3–4 ug/cm^2.

Acute and chronic experiments with implantation of vascular ftorlon-lavsane prostheses (diameter 12 mm, length 35–40 mm) modified by heparin and trypsin into the thoracic section of caudal vein in the dog (i.e. to the site with the lowest rate of blood flow and therefore extremely prone to thrombi formation) corroborate the high hemocompatibility of such polymers. Prostheses retained their permeability for more than 7 months after implantation. In control prostheses, thrombi were found 6–8 hours after implantation in the same location.

One more characteristic property of polymers modified by trypsin and heparin should be pointed out, namely their ability to lyse primary thrombi.

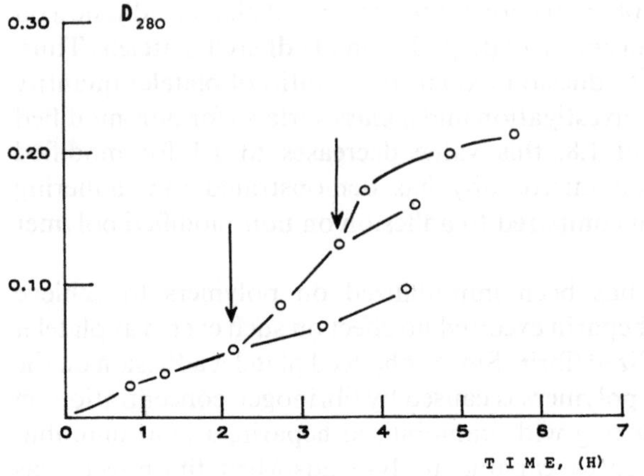

Fig. 40. The kinetics of stabilized fibrin hydrolysis (as OD_{280} increase) by polymers, containing immobilized trypsin and heparin. *Arrows* show the addition of polymer, containing only immobilized heparin [466]

It is well-known that heparin alone (as well as its complexes with various proteins) is uncapable to hydrolyse stabilized fibrin. At the same time, immobilized trypsin retains a sufficiently high activity towards proteins. Besides, the efficacy of fibrin lysis by coimmobilized trypsin and heparin is significantly higher when compared with polymers containing the same (or even higher) trypsin concentration.

To elucidate the reasons for the observed increase in lytic efficacy in the presence of immobilized heparin, polymers containing coimmobilized trypsin and heparin were added to a stabilized fibrin clot. Then, the reaction mixture was supplemented with polymers containing immobilized heparin only. Figure 40 demonstrates the observed kinetics of stabilized fibrin lysis. It is obvious that with each new addition of the heparin-containing polymer, the rate of lysis increases. It has been assumed that this effect is due to complex formation between immobilized heparin and products of fibrin lysis. This may lead to a shift in equilibrium of the lytic reaction towards fibrin degradation.

Polymers with immobilized macromonomers of heparin and trypsin have some advantages when compared with one-component immobilized systems: an increased lytic ability of immobilized trypsin and a decrease in platelet adhesion by immobilized heparin. Such polymers effectively prevent thrombus formation; if this occurs, they can lyse the formed fibrin clot, even when stabilized.

4.3 *Immobilization Principles and Drug Targeting*

4.3.1 *Aims and Tasks of Drug Targeting*

Normally, drugs are more or less evenly distributed in the body and only their activity is expressed maximally in the affected zone. Besides, to reach this zone, the drug must cross many barriers—organs, cells, intracellular compartments, etc., where it can be inactivated. As a consequence, to obtain a high local therapeutic concentration of a drug, one has to administer it in large quantities, the majority of which is wasted by metabolizing in normal tissues. Additionally, in cases of cytotoxic agents and highly antigenic protein preparations, a general negative effect on the organism is gradually developing. To avoid these complications, new methods of drug administration have been suggested, which often are inconvenient for both the patient and the physician (e.g., local infusion of thrombolytic enzymes by means of i.v. catheterization).

The ideal solution of this problem may be drug targeting. Let us define "drug targeting" as the ability of the appropriate drug to accumulate in the affected organ, fast, selectively and quantitatively, independently of the place and the method of its administration. The following advantages of drug targeting are clear: (1) simplification of administration methods, (2) a drastic decrease of drug quantity in the course of therapy, (3) the possibility of increasing local therapeutic concentrations of a drug, and (4) the absence of systemic side effects.

The general ideas concerning drug targeting were formulated at the beginning of our century by Paul Ehrlich, who proposed the hypothetical concept of the "magic bullet". The bullet should be able to find the affected place in the organism and treat it. It is evident that the "magic bullet" has to consist of at least two components: the first should recognize and bind to the target; the second should perform the therapeutic action. Thus, drug targeting in the organism includes the coordinated behavior of three elements: the target, the transporting medium (we shall call it the "vector"), and the drug.

The first step in the solution of this task is the proper choice of a target. It is evident that such a target is a specific organ or tissue. The recognition of the target can proceed on three different levels: on the level of a whole organ, on the level of specific cells characteristic of the target organ, or on the level of individual organelles characteristic of these cells. The most universal form is the recognition of the target on the molecular level. It is well known that every organ or tissue is characterized by a unique set of chemical compounds, among which compounds that are specific for only the given organ are found. In this case, another compound can be used as a "vector" which is capable of specific interaction with the target component (e.g., in forming an antibody-antigen complex).

4.3.2 Principal Systems of Drug Targeting

At present, one can define several possible major schemes of drug targeting [471, 472].

A Direct Application

The direct injection of the native drug into an affected place can also be considered as drug targeting. We can cite as an example the intra-articular administration of hormonal preparations in the therapy of arthritis [473] or the intracoronary infusion of thrombolytic enzymes in the therapy of coronary thrombosis [159, 474]. A very interesting approach has been described by McLaughlin and Goldberg [475], who suggested the intratumor injection of an immobilized antitumor drug—Sephadex-bound adriamycin. Sephadex, activated by periodate oxidation, can bind up to 33 wt% of adriamycin (both covalent and noncovalent). After injection, the preparation can be immobilized in the tumor tissue physically or/and chemically via free aldehyde groups and can release adriamycin by simple diffusion (adsorbed drug) or as a result of biodegradation of the granules (chemically bound drug).

It is evident, however, that this approach can be used only in a limited number of cases.

B Physical Targeting

Physical factors which cause drug targeting can be of both endogenic and exogenic origin. In the first case, the targeted transport can be based on the fact that the affected organ differs in some parameters (e.g., temperature or pH) form the adjacent normal tissues. The most impressive experiments of this kind were performed with the use of liposomal carriers. One study was based on the following [476]: first, inflamed areas are characterized by local hyperthermia; second, it is known that liposome permeability drastically increases at temperatures close to the transition temperature of liposomal phospholipids. Therefore, phospholipids were introduced in the liposome membrane with the transition temperature slightly higher than the physiological one. It was shown for neomycin-containing liposomes that the rate of drug release at 44 °C was 100 times higher than at 37 °C if liposomes were made of depalmitoyl phosphatidylcholine (transition temperature 41°) with the addition of distearoyl phosphatidylcholine. It is evident that local hyperthermia can be also achieved by application of external heat to the target zone, e.g., with microwaves. The same authors [477] have shown that temperature-sensitive liposomes, with incorporated methotrexate injected intravenously into mice with inoculated tumors, released the drug within the tumor several times faster if the tumor was subjected to local external heating. Heat-sensitive immunoliposomes were obtained by Sullivan and Huang [478] by immobilization of palmitoylated monoclonal anti-H2K on the surface of dipalmitoyl-phosphatidylcholine liposomes. The liposomes can bind target cells and release their content after brief heating at 41 °C.

In an other case [479], pH-sensitive phospholipids have been introduced into the liposomal membrane, in view of the fact that inflamed or neoplastic areas usually demonstrate pronounced acidosis. It was shown in experiments in vitro that the incorporation of pH-sensitive palmitoyl-homocystein into the liposome membrane permits the fast and quantitative release of a model compound, carboxyfluorescein, from liposomes at temperatures close to physiological, if the pH is decreased from 7.4 to 6.0. Further development of this approach resulted in pH-sensitive liposomes containing phosphatidylethanolamine and palmitoyl-homocysteine which easily fused during lowering the pH from 7 to 5 with the release of entrapped substances [480]. Inclusion of 40% mol cholesterol into liposomes decreases leakage, but does not influence fusion, which facilitates opportunities for controlled drug delivery. Another kind of pH-sensitive liposomes was obtained utilizing N-succinyldioleoyl-phosphatidylethanolamine [481].

pH-sensitive liposomes can also mediate the cytoplasmic delivery of encapsulated macromolecules. As shown in [482], negatively charged liposomes are endocytosed by the coated vesicle system and accumulate in acidic intracellular vesicles. In this case, liposomes which are unstable at acidic pH (below 7.0) facilitate the cytoplasmic delivery of liposome-entrapped membrane-impermeable macromolecules. Control liposomes do not release entrapped compounds at pH 4–8.

C Magnetic Targeting

An additional example of drug targeting under external physical influences is the use of an external magnetic field to concentrate a drug in the repaired region of the vascular bed. The administered drug had been immobilized on a magnetically sensitive carrier. Some examples of this approach will be discussed in a separate chapter.

D Immobilization of Drugs in Slow Release Systems

Sometimes, the application of established immobilization methods for protein and non-protein drugs on different polymeric carriers [483] leads to dramatic changes in the normal distribution of certain preparations in the body and results in their predominant accumulation in specific organs or tissues. Thus, it is well known that molecules of some synthetic polymers possess a definite affinity for some organs. For example, they can accumulate in the kidney tubules (if their molecular weight is high enough) or be adsorbed on the surface of certain cells with subsequent endocytosis [484, 485].

E Vector Molecules

The most universal method to impart affinity towards target organs or tissues to a drug which, in the normal state, does not possess this affinity is the binding of the drug to a carrier molecule ("vector"), capable of recognizing and binding to the target zone. Here, we will discuss only a few compounds which can be used as "vectors" for drug targeting; exhaustive reviews can be found elsewhere [472, 486]. Let us just mention that among these compounds, there are immunoglobulins, proteins, enzymes, lectines, inhibitors, vitamins, hormones, sugars, etc.

The direct coupling of drug and vector molecules seems to be the simplest way to solve the problem of high efficiency for such systems: if too many molecules will be bound to the vector, the affinity of the latter towards the receptor component of the target organ may decrease sharply.

Nevertheless, in some cases the simplest approach turned out to be quite effective. In the chemotherapy of malignant diseases, direct coupling of the appropriate toxin with antibody against the neoplastic cell led to the generation of an immunotoxin [487]. Immunotoxins are effective due to the extreme activity of the toxin itself, because even a single toxin molecule entering a cell can kill it. The scheme of immunotoxin synthesis based on the plant toxin ricine consists of splitting the toxin into catalytic and binding subunits and subsequent chemical binding of the catalytic subunit to the appropriate antibody. As a result, the toxin acquires the ability to kill selectively only the cells which are recognized by the specific antibody. The biochemistry and application of immunotoxins are an independent field which will not be considered here (see recent publications [488, 489]).

One more example of direct drug/vector binding is the use of estrogen derivatives as vectors to which chemotherapeutic agents have been coupled [490, 491]. Some of these conjugates have shown affinity for estrogen receptors in human breast cancer cells and the ability to kill these cells in cultures in vitro and in experimental animals.

The direct coupling of a drug with vector molecules has also been used for the targeted transport of antiviral preparations into infected cells [492]. Increase in the efficacy of adenine arabinoside, used for the treatment of hepatitis B, was achieved via binding with glycoproteins, containing terminal galactosyl groups. This leads to the increased capture of drug canjugates by the affected hepatocytes, on the surface of which galactose receptors are located.

F Pharmacologically Active Polymers

More sophisticated and more effective systems of drug targeting are known as "pharmacologically active polymers" [33, 493] or "polymeric affinity drugs" [491, 495]. They consist of a biocompatible and biodegradable soluble polymeric matrix, to which molecules of a drug are bound. Simultaneously, the affinity molecule (for example, an antibody) is also bound to the same polymeric matrix, thus performing the recognition function [496] (see Fig. 41).

Several examples of the successful experimental use of such systems have already been described. Goldberg [494, 495] suggested the possibility of regressing atheroscletoric deposits with a polymeric drug consisting of a soluble polymer to which cholesterol oxidase, trypsin, and fibrinolysin are attached together with antibody against the protein component of the atherogenic low-density lipoproteins. In the same papers, a polymeric drug for the treatment of urolithiasis is described, consisting of a soluble polymer with chemically bound hyaluronidase, proteases, and antibody against the protein component of urinary bladder stones.

Some other examples of polymer-drug conjugates are given by Arnon and Hurwitz [497]. It was also shown that daunomycin, bound to monoclonal antibodies to rat α-fetoprotein via a dextran bridge, had a more specific effect on tumors both in vitro and in vivo than the control preparation [498].

In principle, to achieve targeting, one can just impart the carrier polymer property for specific capture by appropriate cells without binding any specific

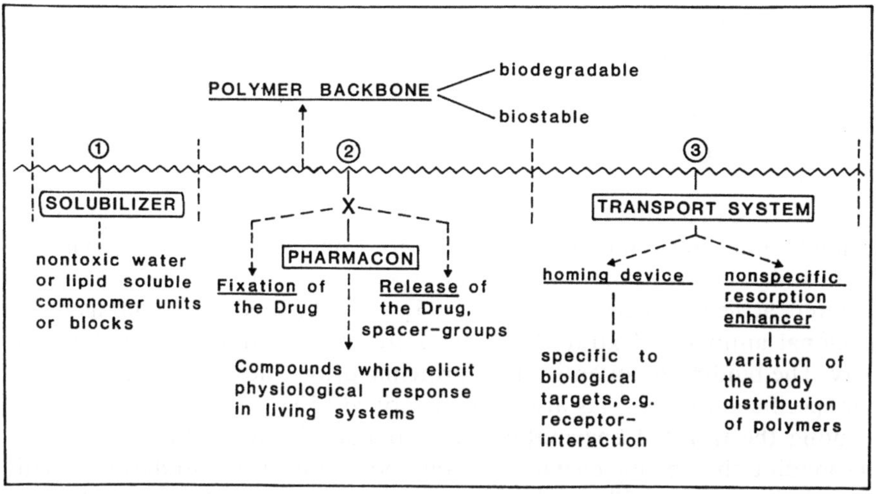

Fig. 41. Targeted polymeric drug (Reproduced with permission from Ringsdorf H (1975) J Polym Sci Symp 51:135)

vector to the polymer. One of the first examples of this kind has been described by Dunkan et al. [499], where terminal galactose residues were incorporated into side chains of poly [N-(2-hydroxyprolyl)methacrylamide]. As a result, the polymer acquired the ability to be captured in the liver by pinocytosis.

G Microreservoir Systems

It has been suggested that instead of a polymeric carrier, a hypothetic microreservoir could be used, whose inner space can be loaded with the drug and the outer surface modified with vector molecules. The drug could be released from the reservoir by diffusional mechanisms or by the gradual destruction of the reservoir under appropriate conditions (as a result, for instance, of the complement-dependent lysis).

What could be used as a microreservoir? First, there are microcapsules prepared from different synthetic or natural materials, which have been developed as drug carriers and exhaustively studied by Chang and co-authors [500]. Second, different cells, notably red blood cells or their ghosts [501], can be used as microreservoirs. Lastly, liposomes can be used as microreservoirs for drug targeting, and to this problem the greatest number of publications in this field is devoted.

4.3.3 Direct and Indirect Conjugation of Therapeutic Enzymes with Specific Vectors

We have already mentioned the studies on the binding of anti-tumor antibiotics with tumor-specific antibodies via a dextran bridge [497, 498]. Many other papers also present interesting data of this kind. The authors of [502] performed covalent binding of methotrexate with tumor, specific antibodies. Investigating the effect of the conjugate, obtained in mice with inoculated tumors, they succeeded to show that mice treated with targeted preparations survived for 90 days, whereas control mice died in 17 days. A comparative study of different preparations of daunomycine, immobilized on anti-tumor antibodies utilizing different methods [503, 504], revealed that the antibiotic directly bound to the immunoglobulin and daunomycine bound to monoclonal antibodies via the dextran bridge possess both a strong and specific anti-tumor activity. Similar results were achieved by the authors of [505], who obtained triple conjugates of monoclonal antibody: dextran T-40: methotrexate and studied their biologic activity. The binding of an anti-tumor proteinic drug—noncarcinostatine with tumor specific antibodies has also been described [506].

Among the targeted enzymatic drugs obtained by immobilization of the corresponding therapeutic enzymes on antibodies, numerous studies deal with thrombolytic enzymes. The right choice of the vector molecule is of great importance in this case. If antibodies are to be used, they should bind specifically with antigens characteristic only for the thrombus. Naturally, fibrin—the main non-cellular thrombus constituent—has been considered as antigen from the very beginning. At the same time, the structural similarity of fibrin and fibrinogen and the presence of common antigenic determinants in their molecules makes it difficult to raise antibodies capable of selective binding with fibrin in the presence of fibrinogen. In vitro, polyclonal antibodies have been used which bound both fibrin and fibrinogen [507, 508]. It was shown that proteolytic enzymes bound to these antibodies via dextran bridges acquire the ability to be selectively adsorbed on artificial fibrin clots and enhance their dissolution.

The authors of [509] performed direct binding of a plasminogen activator—urokinase—with polyclonal rabbit antibodies against human fibrinogen via the heterobifunctional reagent N-succinimidyl 3(2-pyridylthio)propionate. The antibody/enzyme conjugate was purified by gel-filtration. It was shown that both components of the conjugate preserved their specific activity. The conjugate also demonstrated increased stability in human plasma in comparison with native urokinase.

Recently, monoclonals have been obtained against synthetic polypeptides imitating amino acid sequences specific for fibrin (in particular, the amino-terminal sequence of the human fibrin chain) [510, 510a]. These monoclonals specifically bind fibrin but not fibrinogen. They preserve their specificity in the presence of fibrinogen in concentrations similar to that of normal human plasma. The authors have shown [511, 511a] that antifibrin antibodies can be conjugated with urokinase via N-succinimidyl 3(2-pyridylthio)propionate without affecting the specific properties of the enzyme or antibody. The fibrinolytic activity of the specific conjugate is about two orders of magnitute higher than that of the native enzyme (or, an equivalent amount of lysis was obtained with 1/250 of the concentration, with respect to urokinase content, of an antifibrin antibody-urokinase conjugate).

Later, the same authors described antibody-enhanced thrombolysis using tissue plasminogen activators cross-linked with antifibrin antibody 59D8 via SPDP [512]. Antibody-activator conjugates proved to be up to 10 times more potent than tissue plasminogen activator alone and demonstrated more selective thrombolysis. In [512a], the same authors succeeded in constructing a recombinant version of the 59D8—tissue plasminogen activator conjugate. For this purpose, the 59D8 heavy chain gene was cloned and combined in the expression vector with the sequence coding for a portion of the antibody constant region and the catalytic chain of plasminogen activator. The combination obtained was transfected into heavy chain variant cells derived from the 59D8 hybridoma. Thus, a novel hybrid protein was produced capable of simultaneous fibrin binding and plasminogen activation.

In the next step [512b], tissue plasminogen activator was modified by the unidirectional crosslinking reagent N-succinimidyl 3(2-pyridyldithio)propionate and coupled to iminothiolane-modified anti-fibrin antibody 59D8 via the formation of disulfide bonds at neutral pH. In a quantitative rabbit thrombolysis model, the activity of the purified conjugate was compared with that of tissue plasminogen activator alone and that of a conjugate between tissue plasminogen activator and a control digoxin-specific monoclonal antibody. After correction for spontaneous lysis, tissue plasminogen activator—59D8 was shown to be 2.8–9.6 times more potent than tissue plasminogen activator alone.

Although antifibrin monoclonals seem to be a very convenient vector for the targeted delivery of thrombolytics, attempts have been made to use other vectors. Thus, urokinase was immobilized on fibrinogen [513, 514] which had been chosen because it is a natural precursor of fibrin and possesses the ability to accumulate in a thrombus. As a result, its concentration (or, more exactly, the concentration of fibrin formed) exceeds that in the blood. Urokinase was immobilized on fibrinogen via the spacer arm derived from a corresponding aliphatic diamine and it was shown that the affinity of the fibrinogen/urokinase conjugate towards the artificial fibrin clot depended on the length of the spacer arm. The maximal rate of fibrinolysis is achieved with a 1,2-dodecamethylenediamine bridge, which exceeds 2.5 times that of urokinase directly immobilized on fibrinogen. The phenomenon observed can be altered by steric hindrances for the conjugated enzyme binding with the high molecular weight insoluble substrate. The modified fibrinogen preserves its ability to incorporate into the fibrin clot and undergoes polymerization under the action of thrombin.

It was also shown that immobilization of urokinase on plasmin heavy chains via disulfide bridges [515, 516] results in the formation of thrombospecific therapeutic enzymes. The hybrid preparation urokinase/plasmin turned out to be 10 times more active in the treatment of experimental pulmonary artery thromboembolism in rabbits than native urokinase.

Another approach to obtain vectors for targeting of thrombolytic enzymes is based on the observation that thrombin accumulates in platelet enriched regions, which are typical for thrombi. Thrombin may be considered as a potential vector for the delivery of thrombolytics to thrombi. Evidently, thrombin should be first modified in order to suppress its own catalytic (clotting) activity. At the same time, its ability to interact specifically with appropriate receptors on platelet surfaces should be preserved [517]. Proteolytic enzymes (α-chymotrypsin), immobilized on partially inactivated thrombin via carbodiimide activation, dissolve the model fibrin clot 6–9 times faster than the native enzyme.

For some therapeutic enzymes, especially for those which are designated for liver treatment, glycosyl groups can be used as vectors. These groups can be chemically introduced into the enzyme molecule and as long as liver cells have specific receptors for the appropriate glycosyl groups, the modified enzyme begins to accumulate in the liver after intravenous administration. Thus, β-glucocerebrosidase, which is used for treatment of, Gaucher's disease, was covalently modified with N^2-[N^2,N^6-bis[3(α-D-manno-pyranosylthio)-propionyl]-L-lysyl]-N^6-[3(α-D-mannopyranosylthio)propionyl]-L-lysine [518]. After intravenous injection of the modified enzyme into experimental rats, its concentration in liver Kupfer cells was 18 times higher than that of the native enzyme.

4.3.4 The use of Microreservoirs for Targeted Drug Transport

As mentioned earlier, microcapsules, various cells and their "ghosts" as well as liposomes can be used as microreservoirs.

Data on the use of red blood cell ghosts as microcontainers for targeted drug transport are available. The use of multiple protein components of the external membrane of erythrocytes offers many opportunities for immobilization of the corresponding vectors, while the high elasticity of erythrocyte membranes permits firm binding with a target via several bonds, even if the total contact between erythrocyte and target is not large.

Red blood cells, carrying antihuman collagen antibody, are capable of binding specifically (400 to 800 times better than control cells) to collagen-coated surfaces, which can, to some extent, model the injured areas in a blood vessel wall, initiating different pathological processes [519, 520]. The method of antibody immobilization on red blood cells with the help of the avidin–biotin system used by Samokhin et al. [519, 520] is of general interest (shown in Fig. 42). In order to obtain red blood cell-anticollagen antibody conjugates, biotin-modified red blood cells were incubated in an avidin solution and then treated with biotin-modified antibody. The noncovalent avidin–biotin interaction is very rapid ($K_a = 7 \times 10^{-7}\,M^{-1}\,sec^{-1}$) and strong ($K_d = 1.3 \times 10^{-15}\,M$).

Many studies in this field have been performed using liposomes as micro-reservoirs for targeted drug delivery. Some data and problems connected with targeted liposome transport are presented in refs. [521–525].

As already mentioned, the natural target for liposomes is the RES. It offers opportunities for the treatment of liver and other RES organ diseases, which is especially important because many hereditary lysosomal diseases affect liver cells [4–6, 310]. The first clinical experiments of this kind gave promising results.

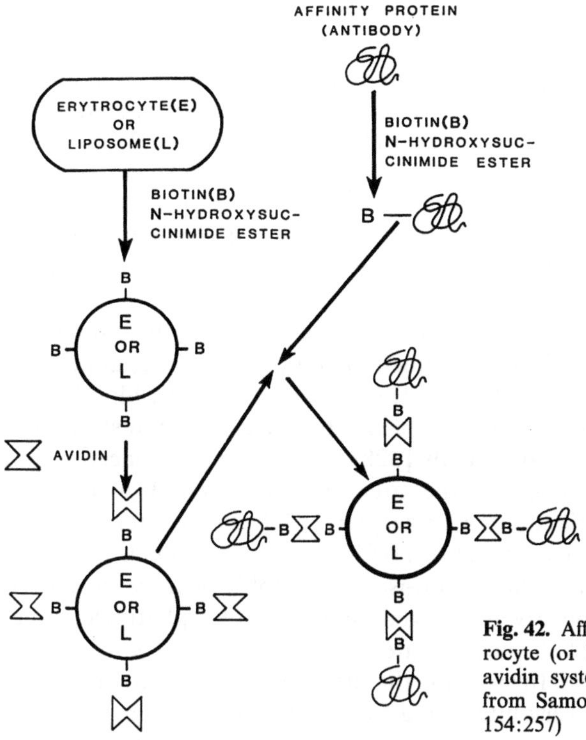

Fig. 42. Affinity protein binding with erythrocyte (or liposome) surface via the biotin-avidin system (Reproduced with permission from Samokhin GP et al. (1983) FEBS Lett 154:257)

Thus, the use of liposomal glycocerebrosidase for the treatment of a patient suffering from Gaucher's disease gave positive results, particularly reduction of the liver size [317]. Administration of the native enzyme had no effect, because it does not enter the cell. A similar result was achieved in the treatment of a patient with glycogenosis type II, when amyglycosidase encapsulated into liposomes was used [288].

The most impressive data are obtained from the experimental treatment of visceral and cutaneous leishmaniasis, which is caused by protozoa locating only in the RES cells [315, 316]. It was shown that the use of highly toxic antimonal drugs in liposomal form permits the drastic decrease of the quantity of the drug used (and, naturally, diminished undesirable side effects) due to the penetration of the drug directly into cells.

Many data exist concerning the possibility of targeting liposomes into different tumors. In some cases, cytotoxic agents incorporated into liposomes cause enhanced tumor regression and longer host survival. However, it is still unclear whether this relates to the actual drug targeting or to the slow drug release from liposomes [490]. Nevertheless, it was shown that the intensification of the therapeutic effect of liposomal illudine was due to specific binding of liposomes to tumor cells [526]. An increase in the efficacy of liposomal doxorubicin action in Ehrlich sarcoma tumor-bearing mice was found by Shinozawa et al. [527].

Rosenberg et al. [528, 529] have shown that spontaneous targeting of liposomes to tumor cells can be achieved with liposomes prepared from autologous phospholipids obtained from the tumor cells. During incubation with ascitic lymphocytic leukemia cells NKLy/LL, liposomes prepared from autologous phospholipids are taken up by the cells two to six times more than control egg lecithin liposomes. Similar results were obtained [530] in liposome uptake by murine hepatocytes in culture, if the liposomes were prepared from murine liver phospholipids and gangliosides.

Changes in liposome composition can sometimes result in their targeting to specific organs or tissues. It was claimed that liposomes prepared from phosphatidycholine, cholesterol and sulfatide (7:2:1) can cross the blood-brain barrier and cause accumulation of liposome-encapsulated horseradish peroxidase in the brain of experimental animals [328a]. The selective concentration of positively charged liposomes in the area of experimental myocardial infarction in animals, described by Caride et al. [318], was later partially confirmed by the data of Kayawake and Kako [531]. These authors have shown that during perfusion of the isolated non-ischemic rabbit heart with a liposome containing perfusate, positively charged liposomes (phosphatidylcholine/stearylamine/cholesterol, 4:1:3) accumulate predominantly. The data for ischemic hearts were equivocal. In any case, the effects observed are too marginal to be of clinical use but are very interesting from a theoretical point of view.

It has been already pointed out that a very general approach to drug targeting by means of liposomes is the immobilization of appropriate specific vector molecules (mainly proteins) on the outer surface of drug-containing liposomes.

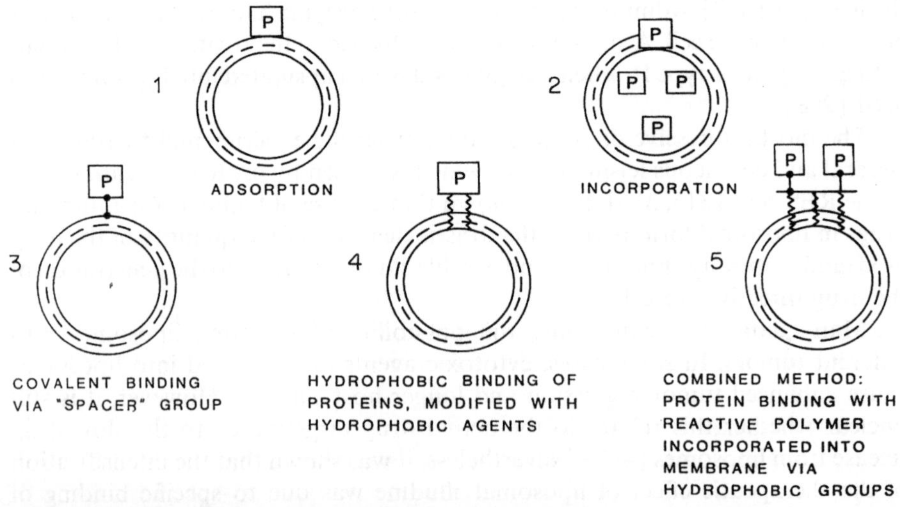

Fig. 43. Possible methods for protein (P) binding with liposomes

It is evident that effective methods of immobilization proteins (such as immunoglobulins) on liposomes must meet a number of special requirements:

1. The specific affinity properties of the vector molecule should remain unchanged after immobilization.
2. The vector molecule should be firmly bound to the liposome.
3. The binding should be efficient and with high yield.
4. A sufficient quantity of vector molecules per each liposome must be bound.
5. The integrity of liposomes during the immobilization should be preserved.
6. The method should be simple enough and suitable for large numbers of affinity molecules.

The following methods can be used for affinity protein immobilization on liposomes (see Fig. 43): adsorption of the protein on the liposomal surface; incorporation of the native protein into the liposome membrane during its formation; covalent binding of a protein to reactive compounds incorporated into the liposome membrane; noncovalent hydrophobic binding of preliminary modified proteins and indirect binding, including covalent coupling of a protein with a polymer. This, in turn, is primarily incorporated into the liposome membrane via hydrophobic residues (which, to some extent, resembles the binding to the cell membrane of a surface membrane protein via an integral membrane protein). This fast developing field in the study of liposomes has been reviewed in [532]. The available data are summarized in Table 17.

A Adsorption

Historically, adsorption was the first method used to bind specific proteins to liposomes.

The most successful example of specific adsorption of a protein on liposomes is the interaction of biotin-containing liposomes with avidin [533]. The approach can be used for the electron–histochemical localization of liposomes in tissues, when the liposome membrane contains a biotin derivative and the sample is treated with a ferritin-avidin conjugate [534].

For the specific sorption of affinity proteins on liposome membranes, other proteins can be immobilized on liposomes before adsorption. Thus, e.g., protein A, which possess high affinity towards immunoglobulins, can be immobilized on liposomes and then any antibody can be adsorbed by these liposomes quantitatively and with high yield [536]. "Second" antibodies can be used instead of protein A; e.g., goat antibodies against rabbit immunoglobulins or antibodies from rabbit sera were used as the "first" specific vectors [536]. This approach permits preservation of the affinity of antibodies, the elimination of steric hindrances for antibody-target interaction due to a long "spacer" arm, and the use of the same initial liposome preparation for binding with different targets by varying the "first" antibodies.

The results of adsorptional immobilization vary in very broad limits from 10^{-5} to 10^{-3} mol of protein per mole of lipid adsorbed.

Table 17. Protein immobilization on liposomes

Protein	Method of immobilization	Binding (mol protein/mol lipid)	Binding yield % initial protein)	Ref.
Immunoglobulin	Protein adsorption on glycolipids-containing liposomes	$5*10^{-9}$	34	[603]
Native or thermoaggregated immunoglobulin	Electrostatic or nonelectrostatic adsorption	$6.7*10^{-5}$	50–98	[604]
Lysozyme	Electrostatic adsorption	$1.6*10^{-2}$	35	[605]
Immunoglobulin fetuin	Sonication of liposome suspension in the presence of protein	$1.8*10^{-5} - 3.1*10^{-5}$	7–28	[594]
Immunoglobulin $F(ab')_2$	Sonication of liposome suspension in the presence of protein	$1.7*10^{-5}$	Not more than 40	[540]
Glycophorin, integral membrane protein	Liposome preparation in protein-containing solution	$1.4*10^{-4} - 2.8*10^{-3}$	30–90	[606]
Immunoglobulin	Covalent binding of carbodiimide-activated protein with liposomes	$3.2*10^{-7}$	12	[542]
α-Chymotrypsin	Covalent binding of protein, with liposomes, preactivated with glutaraldehyde or dimethylsuberimidate	$7.1*10^{-5}$	1–3	[544]
F(ab′)—fragments of antimyosin antibodies	Covalent binding via glutaraldehyde	$8*10^{-5} - 2*10^{-4}$	60	[545]
Peroxidase with protected amino groups	Covalent binding of aldehyde groups of periodate-activated protein with amino groups on liposome surface with subsequent reduction with $NaBH_4$	$1.6*10^{-3} - 6.3*10^{-3}$	Not more than 10	[548]
Antibodies	Covalent binding of protein with aldehyde-containing liposomes	$1.3*10^{-3}$	20	[550]

Monoclonal antibody	Protein treatment with N-hydroxysuccinimide ester of palmitic acid, then liposomes are formed in the presence of protein	$1.6*10^{-4}$	70	[567]
	Protein treatment with N-hydroxysuccinimide ester of palmitic acid, then liposomes are formed in the presence of protein	$5*10^{-4}$	100	[583]
Monomer of Bence-Jones protein	Protein modification with iodoacetyl-dansyllysylphosphatidylethanolamide, then incubation of protein with liposomes	$1*10^{-2}$ ($2*10^{-3}$ from experimental data)	100	[560]
F(ab′)—fragment of antilactose IgG	Protein modification with iodoacetyldansyllysylphosphatidylethanolamine, then incubation of protein with liposomes	$2*10^{3}$	30	[561]
Monoclonal antibody	Covalent binding via SPDP; antibody is bound with protein A on liposome surface	$5.7*10^{-4}$	9–45	[552]
$F(ab')_2$—fragment of antibodies	Covalent binding via SPDP; antibody is bound with protein A on liposome surface	$1.2*10^{-2}$	14–25	[553]
	Covalent binding via SMPB	$1.1*10^{-2}$	20–30	[554]
	Protein covalent modification with phosphatidylethanolamine, then incubation with liposomes	$5*10^{-3}$	45	[566]
Avidin	Protein covalent binding with aldehyde-containing liposomes (oxidized gangliosides)	$7*10^{-4}$	30	[533]
Monoclonal antibody	Protein treatment with N-OSU-ester of palmitic acid, then addition to liposomes obtained by reverse-phase evaporation with subsequent dialysis	$1*10^{-3}$	80–100	[570]
Albumin	Dansylchloride-treated protein is incubated with liposomes	$8.3*10^{-4}$	65	[547]

B Incorporation

The incorporation of proteins into the liposome membrane during liposome formation has been mainly used for reconstitution experiments with integral membrane proteins such as ATPase, cytochrome oxidase, Ca^{2+}-ATPase of sarcoplasmic reticulum, etc. [537–539]. Liposomes were obtained mainly by detergent dialysis, according to which a protein is added to mixed micellae of lipid and detergent.

Some hydrophilic proteins (e.g., β-galactosidase) can not be bound to liposomes by this method [540]. Another drawback of the method is the formation of vesicular aggregates which can be easily removed from the circulation by the RES.

Sometimes, sonication can cause loss of the specific protein activity, as shown by Neerunjun et al. [541] for antibodies against C3HED cells. This may be due to partial protein denaturation during sonication or to the appearance of steric hindrances for the antibody–antigen interaction.

The data presented in Table 17 show that the results obtained are highly variable.

C Covalent Binding

Chemical methods of protein binding with liposomes are based on the immobilization technique, and either the surface of the liposome or the protein molecule (or both) should be chemically activated. Binding can be performed directly or via spacer groups. The main disadvantage of the method (especially in the case of bifunctional reagents) is the formation of cross-links between liposomes or proteins with the formation of large aggregates.

The first attempt to covalently bind proteins with preformed liposomes was made by Dunnick et al. [542], in which carboxylic groups of a specific IgG molecule against thyroglobulin were activated with water-soluble carbodiimide and the activated protein was added to a suspension of preformed phosphatidylethanolamine-containing liposomes. Unbound protein was separated by gel filtration.

A similar attempt with carbodiimide for antibody binding to liposomes was made by Endoh et al. [543]. The degree of binding was 1.3×10^{-4} mol of protein per mol of lipid. The authors point to the formation of cross-linked aggregates of antibodies and liposomes and to pronounced inactivation of antibodies.

It was shown in several experiments [544–546] that the activation of surface amino groups of phosphatidyl ethanolamine in liposomal membranes with dimethyl suberimidate or glutaraldehyde leads to immobilization of the protein (α-chymotrypsin or specific IgG as model) on the liposomal surface via a spacer group during incubation with preactivated liposomes. The method permits almost complete preservation of the specific properties of the protein due to the absence of a direct influence of the liposomal membrane on the bound protein. Thus, 80% of bound α-chymotrypsin retains the ability to be inhibited by a high molecular weight protein inhibitor. The method permits up to 7×10^{-5} mol protein to be bound per mol of lipid. Similar data were obtained with antibodies to canine cardiac myosin (binding degree up to 2×10^{-4} mol protein per mol of lipid (see also Table 17). The total binding yield was about 60%; 72% of bound antibodies remained active in the immunoprecipitation reaction.

Budker et al. [547] used N,N,N′-tris(2-chloroethyl)-N′-(formylphenyl)-propylene-1,3-diamine which was conjugated with phosphatidic acid to produce a low-reactive alkylating agent. The agent was incorporated in liposomes and reduced by sodium borohydride, which resulted in an increase of its reactivity. Activated liposomes bound ribonuclease or hemoglobulin. The relatively low degree of binding (1×10^{-4} mol protein per mol of lipid) can be explained by the existence of a side reaction—activated derivative hydrolysis.

Another approach, involving the chemical activation of carbohydrate residues, was suggested by Heath et al. [548]. Horseradish peroxidase was activated by incubation with periodate and then coupled with preformed phosphatidylethanolamine-, phosphatidylserine-, and stearylamine-containing liposomes. The Shiff bases formed were reduced with sodium borohydride. Binding yield was relatively low (< 10%), but the degree of binding was very high—ca. 6×10^{-3} mol protein per mol of lipid for small monolammelar liposomes. As a result, about 20% of the liposome surface was coated with protein. The loss in enzyme activity was 60%.

A general method for the covalent attachment of immunoglobulins to the outer layer of liposomal membranes has been suggested [549]. Binding also proceeds via aldehyde groups generated in the protein molecule by mild oxidation with periodate on by the action of galactose oxidase. The oxidized protein is then reacted with a hydrazide group linked to a membrane. Using human and murine monoclonal IgM antibodies, the authors showed binding of 535 μg of the protein per μmole of phospholipid. Under optimal conditions, the yield of immobilized protein is 61% in case of small unilamellar liposomes.

The reverse method—periodate activation of glycolipid-containing liposomes and subsequent binding of protein NH_2-groups to activated liposomes—was used by a number of authors [533, 550, 551]. Urdal and Hakomori [533] applied reductive amination for the noninactivating binding of avidin to liposomes.

Direct binding of IgG or $F(ab)_2$ fragments with activated, glycolipid-containing liposomes via aldehyde groups was achieved by Heath et al. [550]. The pH used (8.4) prevented periodate penetration inside the liposomes. Binding was 20%, and the degree binding 2×10^{-3} mol protein per mol of lipid.

With the appearance of the heterobifunctional reagent, N-hydroxysuccinimidyl 3-(2-pyridyldithio)propionate (SPDP), in laboratory practice it was found that it can be used for protein binding to membranes (protein A or mouse monoclonal antibody against human β_2-microglobulin) [535, 522] (see Fig. 44). No aggregates are formed in this process. Barbet et al. [552] succeeded in binding up to 40% antibodies with liposomes. The binding degree was one to ten protein molecules per one liposome with a mean diameter of 58 nm.

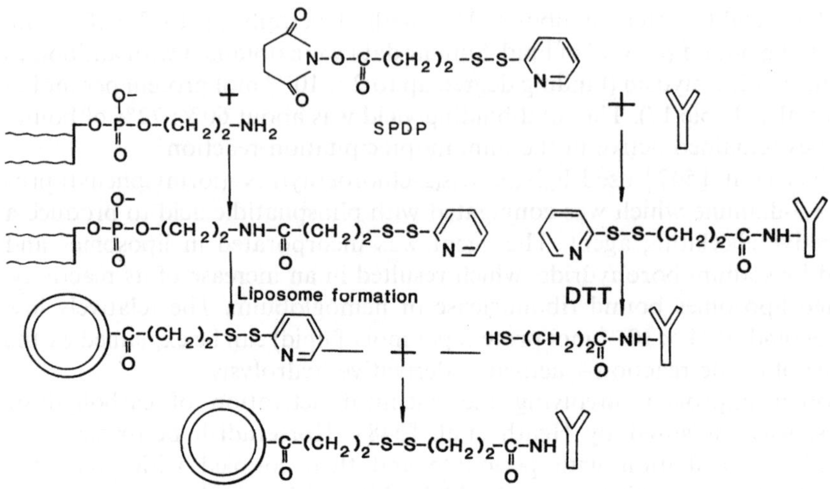

Fig. 44. Protein binding with liposome surface via SPDP

Impressive results with the use of SPDP technique have been reported by Martin et al. (553), who prepared large unilamelar liposomes (0.2 μm in diameter), containing 5 mol% of the dithiopyridyl derivative of phosphatidylethanolamine and then incubated them with Fab' fragments of antibodies against human red blood cells for 2 hr. The proteoliposomes obtained contained about 600 Fab′ molecules per vesicle and were stable in 25% serum. Martin and Papahadjopoulos [554] describe a method for Fab′ immobilization on liposomes with another reagent, succinimidyl-4(n-maleimidophenyl)butyrate (SMPB), which forms nonsplittable bonds. The binding degree in this case was about 250 μg of Fab' per micromol of phospholipid and the protein preserved its specific binding ability. In yet another case [555], SH-bearing subunits of IgM antibody were coupled through the Fc-portion to sonicated liposomes containing N-(m-maleimidobenzoyl)dipalmitoyl phosphatidylethanolamine.

Further development of the method described lead to the introduction of a new heterobifunctional reagent: succinimidyl-S-acetylthioacetate (SATA), as described in [556]. The method is based on the reaction between proteins thiolated with SATA and liposomal maleimido-4-(p-phenylbutyryl) phosphatidyl ethanolamine. The main advantage of this procedure is that it can be used with proteins which do not contain native SH-groups.

Still another method of protein covalent immobilization on liposomes has been described [557]: the inclusion of N-(p-aminophenyl)stearylamide into the membrane produced liposomes that could be activated by diazotization with $NaNO_2$ and then coupled with protein. Approximately 40 μg of albumin per 1 μmol of phospholipid were coupled with liposomes made of phosphatidylcholine, cholesterol and N-(p-aminophenyl)stearylamide at a molar ratio of 15:7.5:1.1.

D Hydrophobic Immobilization of Modified Proteins

It is known that integral membrane proteins can be bound easily and with high yield to liposomes, due to the presence of large hydrophobic domains in these proteins [558]; this makes protein binding to the membrane advantageous from a thermodynamic point of view [559]. Therefore, it was suggested that modification of a non-membrane hydrophilic protein (i.e. immunoglobulin) with a hydrophobic reagent can increase the affinity of the modified protein towards the liposome. Protein binding via the anchor incorporated into the membrane must be firm—ΔG of the transfer of single CH_2 group from water into oil is 0.7 kcal/M, and the equivalent number of CH_2 groups in an anchor group can reach dozens. One of the first examples of this kind was the modification of monomeric Bence–Jones protein with N(N-iodoacetyl-N-dansyllysyl)phosphatidyl ethanolamine, which led to almost quantitative binding of modified proteins with liposomes during joint incubation [560]. Later, the same method was used for the binding of Fab′ fragments of antilactase antibodies to liposomes [561]. Reactive dipalmitoyl phosphatidylcholine derivatives were used to modify free SH-groups in the Fab′ fragment and the incubation of large unilamellar liposomes with 5 μM of the modified protein at a lipid concentration of 1 to 5 μM led to binding of 10 to 20% of the protein added to liposomes.

One more approach for immobilization of modified proteins on liposomes is based on modification of proteins with the reactive derivative of a long-chain fatty acid [562, 563]. α-Chymotrypsin was used as a model protein. NH_2-groups of lysine residues of protein molecules were modified with palmitic acid chloroanhydride (see Fig. 45). The modified enzyme binds easily to liposomes during the process of liposome formation by the cholate dialysis method. The modification of only 3 of the 15 accessible NH_2-groups leads to a binding degree up to 3×10^{-4} mol protein per mole of liquid (15 times more than in case of the native α-chymotrypsin) [564].

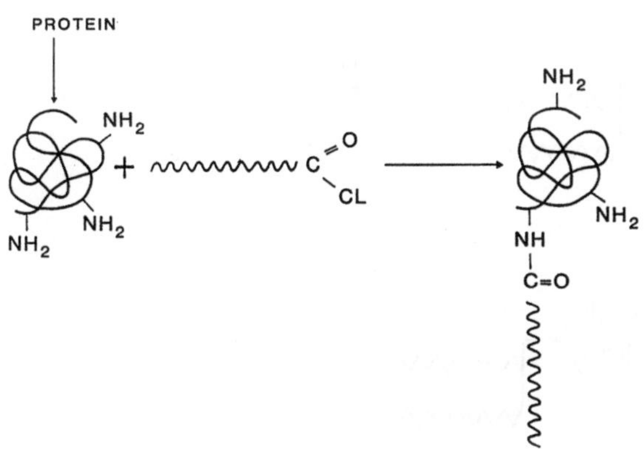

Fig. 45. Protein modification with palmitoyl chloride

A reactive derivative of a natural phospolipid can be used as a modifier. Modification of one amino acid residue in the protein resulted in the introduction of two fatty acid chains in the protein globule phosphatidylinositol (activated by periodate oxidation of its carbohydrate) unit [565] the aldehyde groups formed can interact with protein amino groups (see Fig. 46). Under optimal conditions, the method permits binding of up to 2.4×10^{-3} mol of active α-chymotrypsin per mol of lipid. This corresponds to the immobilization of about 250 protein molecules on the surface of liposome with a mean diameter of about 1000 Å.

$F(ab')_2$ fragments were bound to liposomes according to the following scheme: protein amino groups were reversibly protected with citraconic anhydride, then the protein was activated with carbodiimide and reacted with phosphatidylethanolamine [566]. The purified product was bound to liposomes during joint incubation. The binding degree was approximately 5×10^{-3} mol

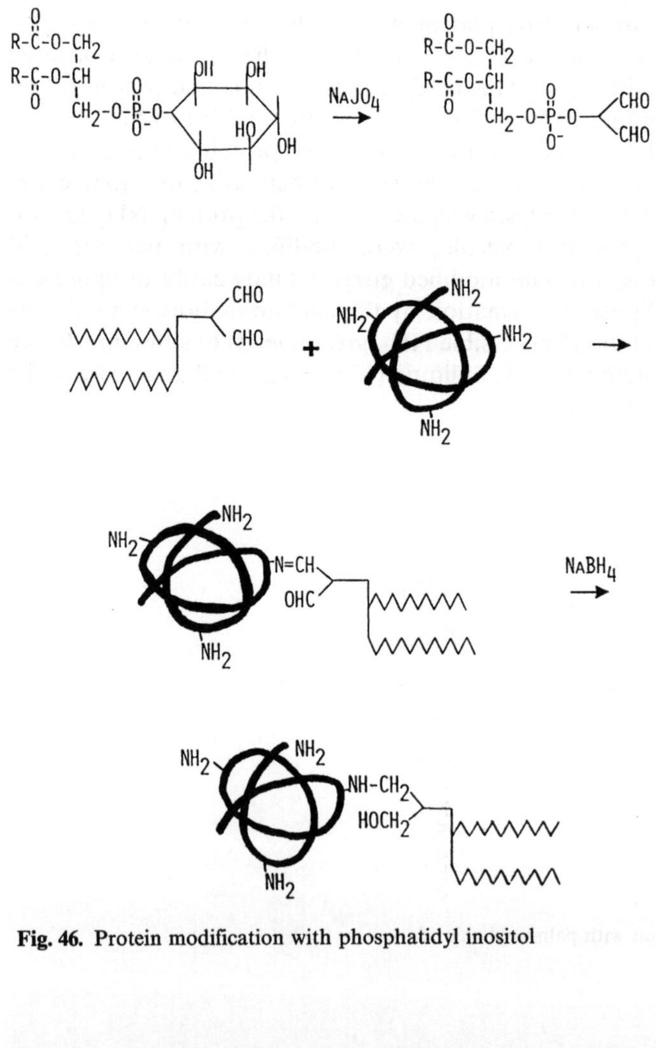

Fig. 46. Protein modification with phosphatidyl inositol

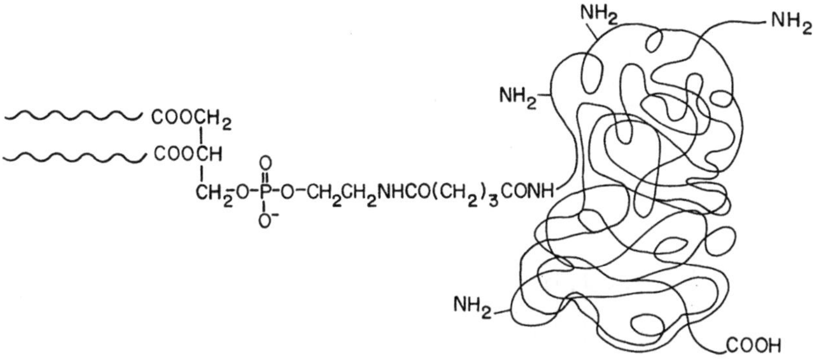

Fig. 47. Schematic representation of proteins modified with *N*-glutarylphosphatidylethanolamine (Reproduced with permission from Weissig V et al. (1986) FEBS Lett 202:86)

protein per mol of lipid. The only questionable point of this method is the citraconilation of 80% of protein amino groups; frequently, such drastic modification can cause protein denaturation. The N-hydroxysuccinimide ester of palmitic acid was used for protein modification by Huang et al. [567–569]. Liposomes were obtained by the cholate dialysis method; the free modifier was separated by gel filtration in a detergent [567]. These authors have also shown that the introduction of four to five palmitic acid residues per IgG molecule gave the best results in terms of preservation of antibody activity binding degree and the ability of the liposome-IgG conjugate to interact with the target.

Palmitoylated antibodies have been incorporated into liposomes by reverse-phase evaporation, when the modified protein was added to the system before complete removal of the organic solvent [570]. This resulted in a binding yield of about 80%.

A new hydrophobic anchor for the immobilization of proteins on liposomal membranes has been suggested [571]. α-Chymotrypsin was covalently modified with N-glutarylphosphatidyl ethanolamine and then incorporated into liposomes during their preparation by reverse-phase evaporation (see Fig. 47). 5.4×10^{-3} mol enzyme/per mol lipid were bound.

E Miscellaneous

Several other methods have been used for immobilization of vector molecules on liposomes. Some are based on the principle shown in Fig. 43, case 5. The suggestion to use polyaminoacids incorporated into the liposome membrane as sites for subsequent protein immobilization was made by Dunnick et al. [542]. We have supported the idea experimentally [572] by incorporating albumin into liposome membranes and subsequently binding another protein (α-chymotrypsin) via glutaraldehyde to albumin. A similar idea was used in [573], where liposomes were modified with commercially available hydrophobic derivatives of different sugars (for instance, with sucrose stearate-palmitate),

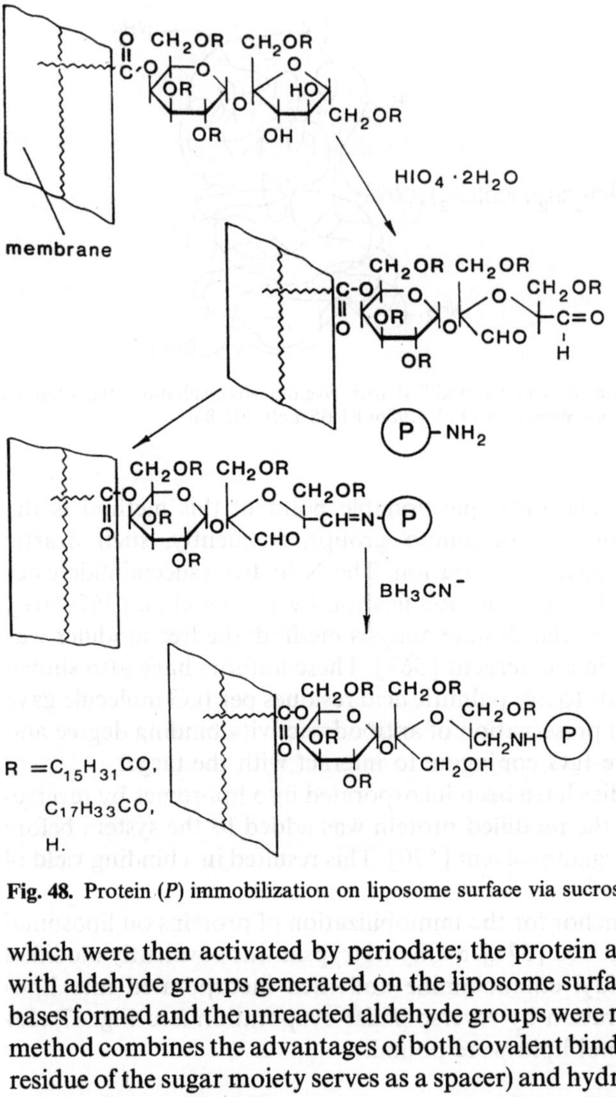

Fig. 48. Protein (*P*) immobilization on liposome surface via sucrose stearate-palmitate

which were then activated by periodate; the protein amino groups were bound with aldehyde groups generated on the liposome surface (see Fig. 48). The Schiff bases formed and the unreacted aldehyde groups were reduced with $NaBH_4$. This method combines the advantages of both covalent binding via a spacer group (the residue of the sugar moiety serves as a spacer) and hydrophobic modification of a protein (fatty acid residues work as anchors fixing the protein on the liposome). The quantity of α-chymotrypsin bound was approx. 10^{-4} mol protein per lipid mol.

Stearic acid-modified dextran also binds to lipid membranes [574] and has been already used for protein immobilization on red blood cells.

F Model Experiments and Quantitation of Specific Liposome Binding

The main requirements which must be met to achieve selective binding of affinity liposomes to a target are:

1. The preservation of vector specific properties upon immobilization.
2. Sufficiently high specific liposome affinity towards a target.
3. The presence of a specific binding site on the target and its accessibility for specific liposomes.
4. Low nonspecific liposome binding with the target.
5. Preservation of liposome integrity during the process of binding with the target.

At least some of the above requirements can be fulfilled in a simple system consisting of liposomes with immobilized specific antibody on their surface and a layer of appropriate antigen.

Important results have been obtained with rabbit immunoglobulin/goat antibodies against rabbit IgG [575, 576]. The antibodies were bound to lecithin liposomes after modification with phosphatidylinositol [565].

To model the naturally occurring situation, when an antigen is a component of insoluble structures (cell surfaces, vessel walls, etc.), a layer of insolubilized antigen was obtained by firm adsorption of rabbit IgG on the surface of a polystyrene dish, similar to a radioimmunoassay or enzyme-linked immunoassay [567].

It was shown that the adsorption of specific liposomes on the antigen layer depended on the concentration of added liposomes and could reach a value that corresponds to complete coating of the antigen (see Fig. 49). At the same time, the nonspecific binding of liposome-antibody conjugates on the albumin layer is negligible. The data in Fig. 49 can be presented in double reverse coordinates or in Scatchard coordinates (Fig. 50), which permits the estimation of the effective

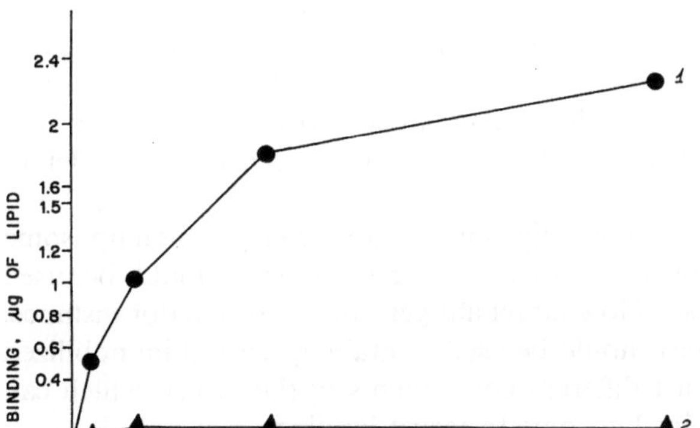

Fig. 49. Binding of lecithin liposomes, modified by goat antirabbit IgG antibodies, with antigen monolayer. (●)—Liposome binding with rabbit IgG monolayer; (▲)—liposome binding with BSA monolayer. Experimental conditions: [^{14}C]—cholesterol labeled liposomes, 0.15 M NaCl, 0.01 M phosphate, pH 7.4; incubation time 1 h (Reproduced with permission from Torchilin VP (1985) CRC Critical Reviews in Therapeutic Drug Carrier Systems 2:65)

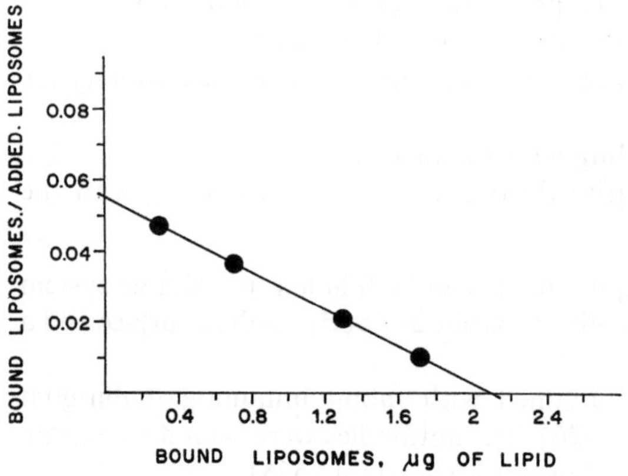

Fig. 50. The presentation of the data in Fig. 49 (for anti-IgG antibody-modified liposomes binding with IgG monolayer) in Scatchard coordinates (Reproduced with permission from Torchilin VP (1985) CRC Critical Reviews in Therapeutic Drug Carrier Systems 2:65)

dissociation of the antigen/liposome-antibody conjugate complex. The value is (1.5 to 5.0) × 10^{-9} M of liposomes (normalized to the 1000 Å liposome containing 10^5 lipid molecules). In a separate experiment, it was established that the dissociation constant of antigen/free antibody complex is 2.5 × 10^{-8} M of antibodies. Comparison of the two values shows that affinity liposome binding to the antigen occurs as multipoint. An similar result was obtained with modified red blood cells [519, 520].

The number of binding sites of the antigen layer is also very important for effective liposome binding. Therefore, to obtain affinity liposomes with monoclonal antibodies, one has to immobilize simultaneously several antibodies against different components of the target on the liposome surface. The increase in liposome size can facilitate multipoint liposome-target interaction, thus increasing the efficacy of the binding. Experimental data of Machy and Leserman [578], showing that small liposomes are more suitable than large liposomes for specific drug delivery in vitro, are in conflict with the above results. Perhaps, this can be explained by the existence of some additional mechanisms involved in affinity liposome interaction with cells, in comparison with the antigen monolayer. Besides, the statement does not necessarily apply to all cell types under all conditions, especially in vivo.

For the fast and effective binding of liposomes with the target, high liposome concentrations and high concentrations of target antigen should be used [545, 579]. In the frequent case of low target antigen concentration (for instance, in tumor cells), large liposomes should be used, containing several immobilized high affinity antibodies against different components of the target, which can provide cooperative multipoint liposome-to-target binding.

G Targeted Liposome Transport to Cells in Vitro

The main body of data on targeted transport of liposomes has been obtained under in vitro conditions. The majority of experiments concern the creation of drug targeting systems for antitumor preparations. Definite success has been achieved in some cell experiments [533, 578] but one should keep in mind that in vivo, tumors are separated from the blood flow by capillary walls. To reach the target cell, the drug-containing liposomes must cross the endothelial barrier, which can turn out to be a problem.

The principal data obtained in vitro in cell experiments are summarized in Table 18. One can see that the results presented vary within wide limits (targeted liposome binding from 20 to 100%). As targets, different cell lines [552, 567], red blood cells [551, 553], affinity columns with immobilized antigen [561] or antigen-coated Pansorbin [566] were used.

One of the first attempts to increase the liposome association with target cells via immunoglobulin adsorption on liposome surfaces was made by Margolis and Dorfman [580]. Carboxyfluorescein-containing liposomes were effectively bound to trinitrophenyl-modified human lymphocytes during joint incubation in the presence of divalent anti-trinitrophenyl antibody [581]; 500 to 15 000 vesicles were bound to each cell. Binding was reduced in the presence of control IgG, in the presence of excess soluble hapten, or in the absence of hapten on cells or liposomes.

The binding of monoclonal antibody to the mouse histocompatible antigen, derivatized with palmitic acid, with liposomes resulted in their specific binding with mouse L-929 cells exceeding nonspecific binding with A-31 cells by 110-fold [567]. The binding was inhibited by free antibodies.

Liposomes bearing mouse monoclonal antibody against human β_2-microglobulin covalently coupled via SPDP bound specifically to human, but not to mouse cells [535]. Covalent binding of antibody to human erythrocyte $F(ab')_2$ fragments with liposomes gives a 200-fold increase in vesicle specific binding with erythrocytes [551]. Using SPDP, anti-β_2-microglobulin antibody could be bound with small unilamellar liposomes. The liposomes obtained bound only to cells containing human β_2-microglobulins as the binding was inhibited by the free antibody in solution [552]. Liposomes, containing methotrexate and carboxyfluorescein, were coupled via SPDP to the monoclonal antibodies against determinants of molecules encoded by the major histocompatibility complex of the murine H-2 haplotype [582]. This resulted in a 75-fold increase in specific liposome binding with blast cells from CBA spleens compared with cells from the B6 strain, which contained no corresponding antigen. Incorporation of palmitoylated monoclonal antibodies against sheep erythrocytes into liposomes led to an 80-fold increase in the binding of these liposomes to cells [583]. Liposomes, coated with subunits of monoclonal IgM antibody against a mouse mammary tumor-associated antigen, also bound selectively to the antigenic target cells [555].

Bragman et al. [584] showed that antibody-targeted liposomes can simultaneously interact with two receptors on K562 cells (a human leukemic line which expresses both glycophorin A and Fc receptors). For interaction with K562 cells, liposomes can be modified with monoclonal antihuman glycophorin A antibodies or with nonspecific monoclonal antibody.

Liposomes with immobilized anti-H2K^k antibody bind with L929 murine fibroblasts 6- to 20-fold more effectively than control liposomes [585]. The

Table 18. Interaction of specific liposomes with targets (in vitro). (From Torchilin, V.P., CRC Crit. Rev. Ther. Drug. Carriers Systems, 2, 65, 1985, with permission)

Liposomes, antibodies, way of antibody binding with liposomes	Target	Ratio: specific/non-specific binding	Bound liposomes (% of total q-ty)	Ref.
$F(ab')_2$ of antibodies against red blood cells, binding with liposomes via gangliosides	Red blood cells	200	80	[549]
Anti-H-2^k-antibodies, binding via palmitoyl-N-hydroxysuccinimidate ester	L-929 cells	300	3	[563]
—"—"—"—"—"—"—"— (larger liposomes)	L-929 cells	30	5–15	[566]
Monoclonal antibodies against red blood cell	Red blood cells	70	11	[579]
Monoclonal anti-H-$2K^k$-antibodies, binding with liposomes via SPDP	L cells, RDM lymphoblasts	20–100	18	[574]
Anti-kappa chain-IgG, binding during cosonication	Lymphocytes, covered with Bence-Jones protein	9	8–62	[575]
Antibodies against concanavalin A	Cells covered with concanavalin A	13	—	[576]
$F(ab')_2$ of goat antibodies against rabbit immunoglobulins, binding via phosphatidyl-ethanolamine	Pansorbin, covered with rabbit IgG	3–5	31	[562]
F(ab') of antibodies against lactose, binding via iodoacetyl-dansyllysylphosphatidylethanolamine	Lactosyl-Sepharose	—	92	[557]
F(ab') of antibodies against red blood cells, binding via SPDP	Red blood cells	40	93–100	[551]
F(ab') of antibodies against red blood cells, binding via SMPB	Red blood cells	20–50	100	[552]
Antibodies against red blood cells, binding via carbodiimide	Red blood cells	9	70	[541]

activity of specific liposome-encapsulated methotrexate-γ-aspartate (which inhibits the growth of L929 cells) increases ten-fold in comparison with the free drug. Machy et al. [586] have shown that antibody-bearing fluorescent liposomes (antibodies bound via SPDP) containing methotrexate were bound to cells expressing the corresponding determinants. Liposomes with anti-H-2K antibodies were more effective for drug delivery into T than B lymphocytes (mainly, via endocytosis). Hashimoto et al. [587] succeeded in increasing the antitumor effect of actinomycin D by entrapping it into liposomes bearing subunits of tumor-specific monoclonal IgM antibody.

Monoclonal anti-Thy_1IgG_1 were used for the targeting of small unilammelar liposomes to mouse AKR-A cells in vitro and in vivo [588]. It was shown that liposome binding to AKR-A cells can be achieved with as little as 1.3 molecules (average) of IgG per vesicle, and after 1 hour incubation approx. 70% of the targeted liposomes associate with target cells. In [589], $F(ab')_2$ fragments derived from anti-rat erythrocyte antibody of normal rabbit serum IgG were immobilized on liposomes made of equimolar amounts of phosphatidylcholine and cholesterol. Targeted liposomes considerably increase their binding to erythrocytes in rat blood, but no increase was observed after their interaction with rat leukocytes or monkey and mouse erythrocytes. In order to obtain specific antiviral preparations, the authors of [590] modified the monoclonal antibody to glycoprotein D of herpes simplex virus and incorporated the modified antibody into liposomes. It was shown that the immunoliposomes obtained bound specifically to infected rabbit corneal cells in vitro.

H Targeted Liposome Transport to Cells and Organs in Vivo

There are only a few examples of targeted liposome transport in vivo. One of the first papers on modified liposome targeting described the tissue distribution of EDTA-containing vesicles in mice [591]. As vector molecules, different glycolipids or brain phospholipids have been incorporated into phosphatidylcholine/cholesterol liposomes. Following the distribution of [^{14}C]-EDTA, it was shown that liver uptake of liposomes with sialic acid or sialogangliosides was lowest. Lungs and brain took up the largest amounts of EDTA from liposomes with sphgingomyelin. Monosialogangliosides and some other vectors caused increased liposome association with spleen and bone marrow. Nevertheless, in general, the majority of liposomes were taken up by the liver (from 23 to 48% of the total).

Phosphatidylcholine/cholesterol liposomes containing aminosugar derivatives as vector molecules in mice administered subcutaneously, concentrated mainly in mice in polymorphonuclear leukocytes [592, 593]. This can be explained by the existence of specific receptors on the target cells.

Many attempts have been made to increase liposome binding for definite types of liver cells—hepatocytes or Kupfer cells. Gregoriadis and Neerunjun [594] found that liposomes coated with asialofetuin were bound more efficiently to liver parenchymal cells than free liposomes. Liposomes coated with β-galactoside were captured by rat hepatocytes three times more efficiently than by nonparenchymal cells [595]. At the same time, liposomes with α-mannoside residues on their surface are taken up preferentially by nonparenchymal cells. Liposomes containing galactolipid have higher affinity to hepatocytes of mouse liver than liposomes without galactolipid [596]. Generally speaking, glycosylated liposomes are captured by the liver much faster than nonglycosylated liposomes. This fact was used for the enhanced delivery of liposome-immobilized β-glucocerebrosidase into liver cells [597]. Spanjer and Scherphof [598] have shown that the incorporation of 8% mol of lactosylceramid in liposomes prepared from cholesterol, dimyristoylphosphatidylcholine, and phosphatidylserin in a molar ratio of 5:4:1 lead to increased capture by the rat liver, mainly due to the enhanced uptake by hepatocytes via a galactose-specific receptor. Lactosylceramide was also used to increase the delivery of liposome-entrapped plasmid containing the preproinsulin I gene into rat liver hepatocytes in vivo [599].

Coating liposomes with polysaccharides can also improve the selectivity in the delivery of liposome-entrapped substances into specific organs. Liposome coating with different polysaccharides (palmitoyl conjugates of pullulan, pullulan phosphate, amylopectin, amylopectin phosphate and amylopectin sulfate) increased liposome uptake in guinea pig lungs 3 to 5 times after intravenous injection [600, 601].

Liposome-antibody conjugates were used for targeted liposome transport into necrotic tissues in dogs with experimental myocardial infarction [545, 546]. Liposomes containing the radioactive marker $^{111}InCl_3$ were covalently coupled via glutaraldehyde with anticanine cardiac myosin antibodies or their Fab fragments. Activity of liposome-immobilized antibodies in the immunoprecipitation reaction towards the soluble antigen was preserved at approx 70%. Rapid accumulation of radioactivity in the necrotic zone after i.v. or intracoronary injection of targeted liposomes was proven by γ-imaging. Nonspecific IgG liposomes or free markers showed no binding in the target area. The suggested approach can be used for both drug transport and for diagnostic purposes.

There still exist many unsolved problems in targeted liposome transport to cells and organs in vivo. One of them is the ability of intravenously injected liposomes to cross vascular barriers. In a detailed review, Poste [602] demonstrated that liposomes are unable to cross continuous or fenestrated capillaries and have limited ability to reach extravascular targets with the exception of tissues containing sinusoidal capillaries.

The use of liposomes in systems of "reversed targeted transport" is very interesting. In this case, the increased uptake of liposomes by RES cells is used for in vivo redistribution of different substances. For example, the excess of diagnostic proteins after radioimmunoscintigraphy can be removed from the blood very rapidly after administration of liposomes with immobilized antibodies against this protein on their surface [607]. A similar attempt was made in [608] in order to remove toxins from the blood, using the system liposome/antibody.

4.3.5 Enzyme Immobilization on Magnetic Carriers

As has been pointed out in Sect. 4.3.2., magnetic derivatives of drugs and their local accumulation in the target tissue under the action of locally applied external magnetic fields are an interesting and promising approach of targeted drug transport in general and targeted enzyme transport in particular.

In principle, magnetic materials have been proposed as carriers for protein immobilization more than 10 years ago [609, 610]. The ability to concentrate near magnetic poles after application of an external magnetic field was considered to be the main merit of such carriers. This ability has been used in technological processes for the selective removal of enzymatic catalysts from reaction mixtures [611], as well as in immunological studies, in the separation of cells preliminarily treated with magnetic microparticles modified with antibodies against specific components of the cell surface [612]. A review of magnetic carrier properties is presented in [611].

Magnetic carriers have been more and more frequently used for the targeted transport of immobilized therapeutics [377, 613–618]. General considerations as well as experimental studies indicate that the following structures of magnetic preparations are possible in principle (see Fig. 51): polymeric microspheres, containing a mixture of fine ferromagnetic particles and drug (enzyme) molecules; in the inner volume magnetic microparticles coated with polymer and containing a drug immobilized on the external surface of the granule; and finally, carriers of "artificial cell" type, containing magnetic particles in the internal water space, the therapeutic drug may be trapped in the internal space of the carrier or immobilized on its external surface or both. Among the materials currently in use, carriers based on polysaccharides [613, 618], albumin [616, 617] or red blood cell membranes [377] are the most popular.

It stands to reason that the use of such carriers raises the problem of the safety of the method. The main problems arising are the following: the danger of small vessel embolization by carriers; the problem of overloading of the reticulo-

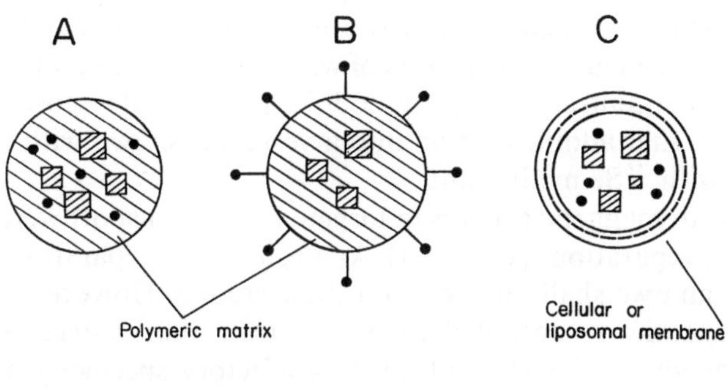

Fig. 51. Possible structures of magnetic carriers for drug (enzyme) immobilization

endothelial system with drug particles; the danger of development of toxicity due to ferromagnetic microscopic particles accumulating in the body (usually Fe_3O_4); the selection of optimal intensity and gradients of the magnetic field, sufficient for retaining drug particles in the blood flow, and, at the same time, exhibiting no undesirable biological effects; minimization of interactions between drug particles, which may cause the formation of aggregates [377].

The problem of hydrodynamics of magnetic particles in a high-viscosity system (blood) has been well elaborated, both experimentally and phenomenologically. The calculations presented in [617] permit to estimate the forces which act on ferromagnetic particles of a given size and magnetite content moving in the blood flow, in a permanent magnetic field of a given gradient. On the other side, based on a mathematical model of magnetic particles behavior in the blood flow and their organ distribution, general conditions have been, formulated for their application in practical medicine [619]. It was shown that this approach is applicable only for drugs with short life span in the blood (a half-life of not more than 5 min) and the ability to cause the desirable biological effects in low concentrations (several ug/ml).

It is generally accepted, based on experimental data and theoretical considerations, that at physiological magnetic field intensities (not more than 40000 Oe) and known rates of blood flow in various vessels, ferromagnetic therapeutic drugs with a particle size $\sim 1\ \mu m$, containing up to 50% (w/w) of magnetite, may be reliably stopped only in small vessels, where the linear blood flow rate is about 0.05 cm/sec (in capilaries and not in vessels where the blood flow rate may reach several cm/sec).

Interesting regularities have been observed in comparative study of the retention of magnetic carriers of various structure and particle size by magnetic fields in experimental animals. After administration of various radiolabeled magnetic carriers into the marginal rabbit ear vein "under a magnet" (the design of the experiment and part of the results, Fig. 52), large particles (magnetized Sephadex granules, 20 μm) and particles of colloidal size (Fe_3O_4 stabilized with polymer, 50 nm) are retained in the blood flow with different efficacy [620, 621]. While large particles are reliably retained even in comparatively large vessels by the field, the same is not true for small particles. However, small particles can be assembled into relatively stable aggregates, the size of which is similar to that of large particles. As a result, a concentrated magnetic colloid of enhanced viscosity moves as a compact "drop" after administration into a vessel (as shown with a magnetic colloid, labeled with ^{75}Se-methionin).

The first experiments with magnetic carriers in targeted drug transport were performed with antitumor preparations [613–616]. As a rule these preparations are low molecular weight and we shall not dwell on these studies. However, it should be noted that the authors succeeded to demonstrate the absence of toxicity from magnetic carriers [617, 621] and quite satisfactory specificity of localization at the site of the magnetic field.

One of the first examples of magnetic preparations of immobilized enzymes for targeted drug transport has been described in [618]. The authors obtained

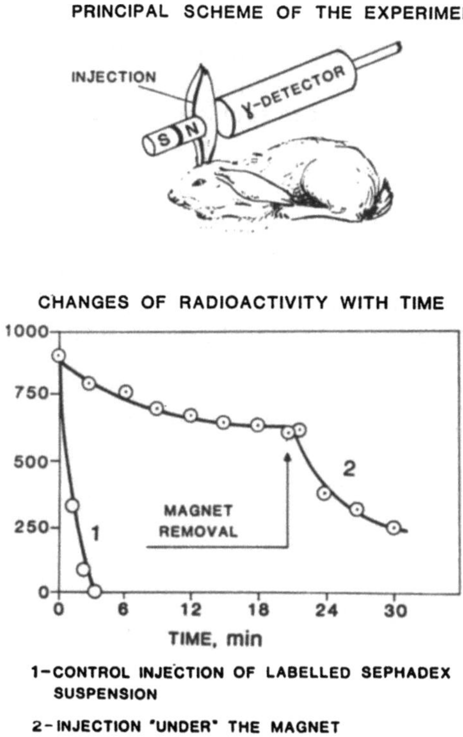

Fig. 52. The scheme of the study on targeted transport of ^{131}I-labeled albumin immobilized on Fe_3O_4-containing aldehyde-Sephadex G-25 in the rabbit ear marginal vein (Reproduced with permission from Torchilin VP (1987) Adv Drug Deliv Rev 1:41)

microspheric magnetite preparations (2–3 μm) coated with starch, on which ^{125}I-labeled albumin and plasmin were immobilized via starch activation with cyanogenbromide. In animal experiments, magnetic preparations were injected into the marginal rabbit ear vein and the magnet was fixed to the other ear of the same animal. The authors succeeded in achieving a certain concentration of the magnetic preparation in the area of the magnet, although the amount of magnetic carrier located in the counterlateral ear represental only a small fraction of the total injected β-D-glycosidase has been immobilized on synthetic polymer coated nickel, cobalt, tin, iron and albumin particles [620]. Strict requirements for biocompatibility and biodegradibility of the polymer coating with immobilized enzymes for medical applications have been formulated [622].

Magnetic Sephadex, into the pores of which ferromagnetic material was sedimented [623], has also been used for enzyme immobilization. Sephadex was activated with cyanogenbromide or by periodate oxidation. The latter method may be performed without the use of highly toxic intermediate compounds. α-Chymotrypsin, immobilized on such carriers, retained almost unchanged catalytic potential for hydrolysis of the low molecular weight specific substrate, N-acetyl-L-tyrosine ethyl ester.

During the last years, studies have emerged both on improvement of already known methods for obtaining magnetic carriers (e.g. made of albumin [624], and on the development of new carriers, e.g., based on calcium alginate [625]. The entrapment of fine magnetic material into liposomes has also been described [626]. In the latter study, the authors succeeded to entrap up to 15 μg magnetite per 1 mM of lipid and demonstrated that liposomes can simultaneously entrap high molecular weight markers into the internal aqueous space. It was also shown that magnetic liposomes are effectively accumulated in the area of magnetic field application, both in vivo and in vitro.

Most recent studies describe the production and application of magnetic preparations of immobilized thrombolytic enzymes [627–629]. These studies demonstrate, both in vivo and in vitro, that an artificial thrombus may be lysed by a magnetic fluid containing thrombolytic enzyme. In model experiments with

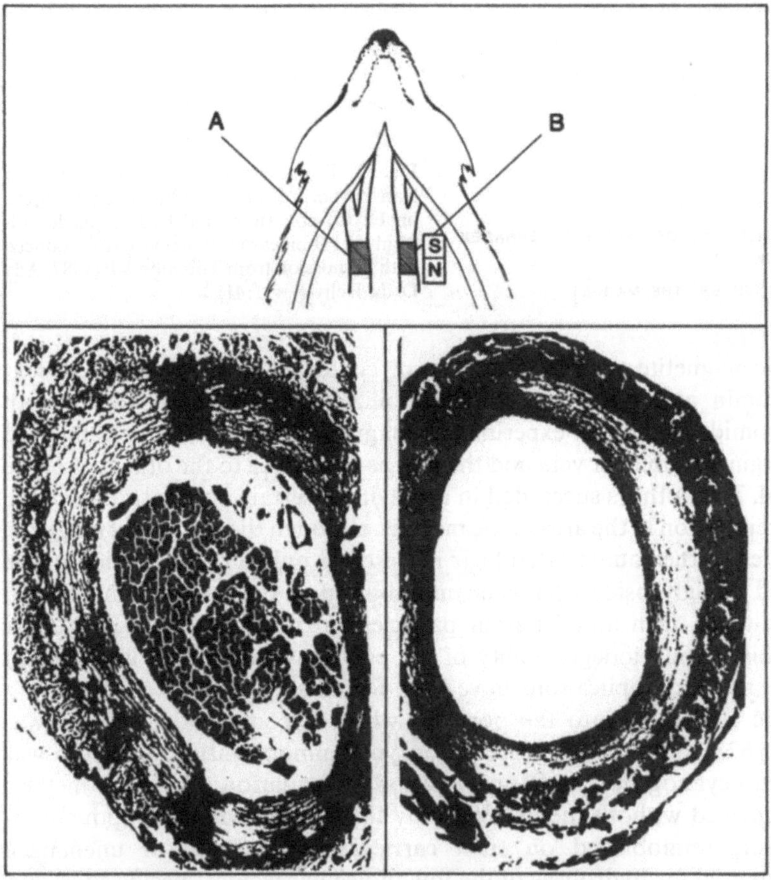

Fig. 53. Experiment on the lysis of an artificial thrombus in the dog carotid artery. **Top**—scheme of experiment: Sm—Co magnet is implanted in the vicinity of thrombus in one artery; **Bottom**—frontal sections of both carotid arteries 6 hours after experiment, left—artery without magnet (occluding thrombus can be seen), right—artery with magnet (thrombus absent)

thrombolysis, one of the vessel branches was completely embolized with a fibrin clot imitating thrombosis [628]. Then, a thrombolytic preparation consisting of plasmin and magnetic fluid was administered via catheter. When a magnetic field was applied (with an intensity of 8000–9000 Oe and a gradient of 15–20 kOe/cm), the preparation was drawn into the embolized branch by the field and the enzyme lysed the fibrin clot.

The possibility of fibrin clot lysis by magnetic sedimentation of microspheres with proteolytic enzymes immobilized on their surface has been demonstrated [629]. Studies were performed with a model of blood vessels (the diameter of the polymeric tubes corresponded to that of small arteries and was 1.5–2 mm). One of the branches was embolized so that the remaining flow constituted less than 1% of the flow without clots. A suspension of microspheres with trypsin was introduced into the solution circulating through the system and retained on the clot surface by a magnetic field. It was shown that only the clot with sedimented microspheres was lysed and soon disappeared.

The relief of the surface on which magnetic carriers are sedimented plays an important role. It was shown microscopically that microspheres were stopped and accumulated mainly in places of low hydrodynamic velocity and shear intensity: in folds and cavities of the microrelief of the clot surface. This fact indicates the possibility of selective deposition of carriers by magnetic fields in embolized vessels with rough microrelief [628, 629].

Recently, the targeted delivery of streptokinase immobilized on a magnetic carrier to the region of artificial thrombus formation and the ability of a targeted drug to perform effective thrombolysis in the desired location [629] has been achieved. In this study, thrombus formation in both dog carotid arteries was induced by interposition of a trapezium-like vessel wall fragment into the arterial lumen. A permanent Sm—Co magnet was implanted into the tissues close to one of the vessels in the region of the thrombus (Fig. 53, 54). The thrombolytic preparation, containing immobilized streptokinase, was administered via the back leg vein in three equal portions (20000–27000 IU each, in 1 ml of water) at

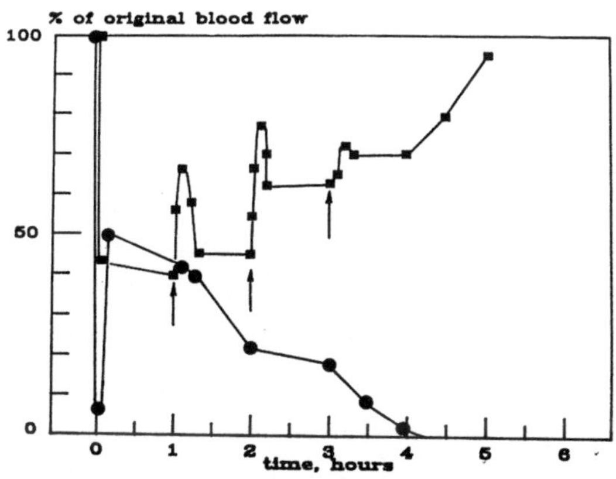

Fig. 54. Blood flow (% from initial) in dog carotid arteries after induced thrombosis. Control artery without magnet (●) artery with implanted Sm—Co magnet (■). *Arrows* indicate time of "magnetic" streptokinase injection

one hour intervals between injections. Immediately after the operation, the blood flow in both arteries decreased drastically by up to 40% of the original value as a result of the spasm caused by surgery. A further gradual decrease in blood flow was caused by the process of thrombus formation. In the control artery (without magnet), the blood flow disappeared completely in all cases 4 hrs after surgery. In the artery with the implanted magnet, gradual restoration of the blood flow began after administration of the preparation and returned practically to normal level after 30 min. The histological analysis of arterial sections obtained 5 hrs after the operation showed the existence of red thrombi in the control arteries with complete vessel occlusion. In the artery with the implanted magnet, thrombus were not observed (Fig. 53). The dose of streptokinase used was much lower (10-fold) than in case of routine systemic application of the enzyme.

The data demonstrate the high efficacy of local thrombolysis (at least in the case of fresh thrombi).

5 Immobilized Enzymes for Extracorporeal Application

Here, we will consider the state-of-the-art in the field of immobilized enzymes which are used either for the treatment of external lesions (wounds, ulcers, burns) or for specific action on biological fluids pumped through temporary extracorporeal equipment.

5.1 Immobilized Enzymes in Extracorporeal Systems

Immobilized enzymes have been applied in many extracorporeal systems, in order to remove undesirable metabolites from biological fluids during perfusion. Due to the variety in construction, immobilized enzymes can be used as membranes, coils, microgranules, tubes etc. The development of extracorporeal apparatus with immobilized enzymes is of primary interest in the treatment of liver failures, blood diseases and cancer (e.g., the removal of asparagine from the blood with immobilized asparaginase in the treatment of asparagine-dependent tumors).

The use of extracorporeal apparatus completely eliminates the contact of immunocompetent cells with foreign proteins, sharply decreasing immunologic and toxic side-effects, and permits the prolonged treatment (continuous or periodic) with decreased cost and enzyme waste. In some cases, minor and easily eliminated complications can be observed—the formation of fibroid films on the inner reactor surface (which is especially true for reactors with high resistance to the blood flow); partial hemolysis and decrease in the number of red blood cells and platelets (despite of rapid restoration of normal values after completion of the procedure). The use of membrane reactors is recommended instead of column reactors with granulated immobilized enzymes; sometimes, non-specific sorption of blood proteins takes place which must be prevented.

Enzyme immobilization on the inner surfaces of a reactor in contact with the blood, increases hemocompatibility and athrombogenicity (by passive action of an enzyme). On the other hand, the effect of an extracorporeal apparatus can be based on an enzyme-catalyzed process, which usually removes the excess of undesirable metabolites from the blood. This approach can be considered as an extracorporeal enzyme replacement therapy. Various enzymes are often coimmobilized with the therapeutic enzyme for better biocompatibility of the whole extracorporeal system. In more complex cases, when the concentration of "pathological" metabolites is very low, the extracorporeal system can additionally possess the ability to concentrate a given metabolite in the reactor, where it is eventually degraded by an immobilized enzyme. For this purpose, an affinity ligand is incorporated in the reactor either independently or coimmobilized with the therapeutic enzyme on the same matrix. Optimal biocompatibility and non-thrombogenicity can be achieved when the layer of enzyme-containing hydrogel is formed on the surface in contact in blood.

Nevertheless, some traditional methods of enzyme immobilization are still in use to obtain materials for extracorporeal vascular shunts. Fibrocollagenic tubes with immobilized urokinase have been described [630–633]. Immobilization was performed by enzyme adsorption and subsequent cross-linking with glutaryc dialdehyde. The authors have shown that after shunting the dog carotid with the fibrocollagenic reactor containing immobilized urokinase, low adhesion of platelets or fibrin has been observed.

The kinetics of fibrocollagen-immobilized urokinase is practically the same as of the native enzyme. The operational stability of the immobilized enzyme can last as long as 50 h. Enzyme-containing tubes can be stored for one year without any loss of urokinase activity.

Thrombolytic enzymes have also been covalently immobilized on the surface of polyurethane. Athrombogenic polyurethane was obtained with urokinase, streptokinase and several non-enzymatic thrombolytic drugs [634].

Urokinase-containing polymers can be used in the manufacturing of vascular prostheses, extracorporeal reactors and shunts [635–640]. Urokinase immobilization on partially hydrolyzed nylon has been described [640]. Basic and clinical studies on urokinase-containing materials derived from nylon, polyesters, polyvinylchloride and polyurethane have been reviewed [639]. Some of these materials have already been used to manufacture intravenous catheters and lymph drainages and demonstrated a high degree of biocompatibility.

Systems with immobilized enzymes can solve different tasks: they can decompose accumulating toxic metabolites in the blood (urea, hydrogen peroxide, superoxide radicals and many others); they can remove from the body toxic exogenous substances from drug overdose (toxins, narcotics, heparin); they can remove substances which are non toxic, but stimulate pathological processes; they can serve as activators of some normal physiological systems in the body (e.g., activation of plasminogen during thrombolytic therapy). The mathematical analysis of the functioning of extracorporeal reactors with immobilized enzymes is not sufficiently developed and so-called "general considerations" are used too often. Studies attempting to develop the pharmacokinetics of extracorporeal reactors have introduced a new parameter—"bioefficiency" (calculated from the combined analysis of reactor behavior and pharmacokinetics of enzymatic reaction products) [642]. The linear pharmacokinetic models suggested for the definition of bioefficiency are based on data on the extracorporeal decrease in metotrexate toxicity by the action of immobilized carboxypeptidase G_1 [643].

Mathematical models of tubular extracorporeal shunts with immobilized enzymes have been analyzed [644]. The authors established the connection between an average concentration of enzymatic reaction products at the reactor exit and reactor parameters. They also formulated limiting kinetic laws for the reactor productivity; their theoretical calculations coincided well with results from experiments performed on extracorporeal shunts with lavsan-immobilized asparaginase.

Many studies are devoted to the use of enzyme-containing reactors in the therapy of substrate-dependent tumors. Thus, asparaginase was immobilized by incorporation into fibers made of cellulose triacetate [645] or dacron [645], covalently bound on the inner surface of nylon tubings [647] or collagen films [648]. As a result, systems have been developed capable of almost quantitative removal of asparagine from the blood, after 0.5–2 hours of perfusion, in dogs and monkeys [648] or even man [649]. Furthermore, this therapy is not accompanied with immunological complications, which are usually observed after administration of the native or even soluble polymer-immobilized enzyme. Data on asparaginase immobilization into hollow fibers used as reactor filler are presented in [650].

The flow kinetics of L-asparaginase, immobilized on the surface of nylon tubings have been studied [651]. Another model of an extracorporeal shunt with asparaginase was developed with a lavsan vascular prosthesis [652]. Asparaginase was immobilized on the prosthesis surface by macromolecular complexing with collagen. Then, the prosthesis was placed into a silicon tube and coiled. When fluid moves through the coil, a convective flow pattern forms, which promotes mixing of the solution; the productivity of the reactor increases with increase of the flow rate. When this shunt is perfused in a closed cycle with one liter of asparagine solution (initial asparagine concentration 10^{-4} M; pH 7.8; 37 °; flow rate 93 ml/min), 95% of the asparagine present is broken down during the first 30 min of perfusion. When whole human blood was perfused through the shunt (total volume 120 ml; 20 °C; flow rate 38 ml/min), 90% of the asparagine present disappeared during the first 20 min of perfusion. No hemolysis or thrombus formation could be noticed; the blood cells remained unaffected. The shunt was still active after two months; contact with blood did not affect its activity.

In [653], L-asparaginase was immobilized in hollow fibers made of porous polypropylene via grafted residues of methacrylic acid. This material was tried in the experimental treatment of acute lymphocytic leukemia; the extracorporeal device with the immobilized enzyme was introduced as a carotid-jugular neck shunt in sheep. The authors demonstrated the high stability of immobilized enzymes in the device and their ability to decrease blood asparagine from 5.2 mM/dl to a level below the sensitivity of the determination method in 45 minutes (Fig. 55).

The first results from clinical trials with extracorporeal devices with immobilized asparaginase have been described in [654]. L-asparagine was immobilized in the reactor with hollow fibers made of regenerated cellulose

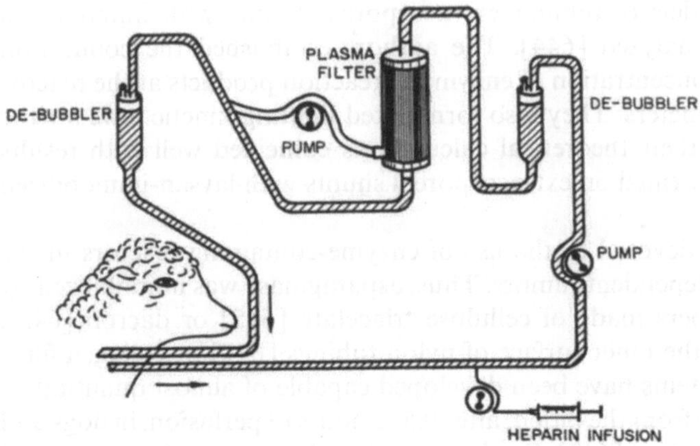

Fig. 55. Schematics of the in vivo testing of a device with immobilized L-asparaginase in sheep (Reproduced with permission from Gombotz WR et al. (1985) J Contr Release 2:375 Copyright 1985, Elsevier Science Publishers B.V.)

(immobilization was performed via trichloro-s-triazine). The total surface area of the reactor was approx. 2.5 m^2. At flow rates of 200 ml/min, the decrease in blood asparagine concentration was 30% higher in a hollow fiber reactor with immobilized asparaginase when compared with traditional dialysis. The use of a hollow fiber reactor in three patients with tumors lead to noticeable temporary improvement in two of them. Furthermore, the use of asparaginase-containing extracorporeal devices did not provoke any immunological reactions even in patients who had already demonstrated anaphylaxis after parenteral administration of the native enzyme.

Extracorporeal reactors with immobilized arginase can be applied in the treatment of arginine-dependent tumors. Arginase was immobilized in hollow fibers [655, 656] and used in the treatment of familial hyperargininemia (in model experiments performed with rabbit blood). In the reactor, arginase was separated from the blood with a membrane which was permeable only for low molecular weight substances, arginine being among them.

In the treatment of phenylalanine-dependent tumors, the enzyme phenylalanine ammonia lyase which converts phenylalanine into non-toxic trans-cinnamic acid may be promising. Therapy with the parenterally administered native enzyme causes frequent immunological complications, especially after repeated injections. This is also true to some extent for the parenteral administration of the enzyme immobilized on soluble carriers. The possibility of using devices with immobilized phenylalanine ammonia lyase has already been experimentally proven [657]. The enzyme, immobilized in a hollow fiber reactor, removed phenylalanine from a sample of citrate blood rapidly and quantitatively. The experimental treatment of phenylketonuria was performed in dogs. Shunting of the dog femoral artery and vein with a multi-tubular device containing phenylalanine ammonia lyase immobilized by incorporation into hollow fibers or by chemical fixation on the inner surface of nylon tubings leads to a 2-fold decrease in blood phenylalanine [658]. Low phenylalanine levels can be maintained during at least two days. Similar studies using fiber reactors with immobilized phenylalanine ammonia lyase for the treatment of phenylketonuria have been described in [659]. The use of an enzyme reactor, shunted between the femoral artery and femoral vein of a dog for two hours, decreased blood phenylalanine levels from 15 to 2 mg/100 ml. No changes in the number of red blood cells, leukocytes and platelets have been observed. Some data concerning immobilized phenylalanine ammonia lyase are also presented in Refs. [660–662].

Kidney insufficiency can also be treated by removal of toxic metabolites from the blood in extracorporeal devices with immobilized enzymes. Excessive urea can be eliminated from the blood in a reactor with immobilized urease. This was proven using a hollow fiber reactor with immobilized enzyme. A complication of this procedure is the formation of ammonia in the process of enzymatic hydrolysis of urea, which also has to be removed from the blood [663]. To remove ammonia, a urease-containing reactor can be combined with a reactor containing an adsorbent, such as activated charcoal. The use of a conjugated column reactor with coimmobilized lactate dehydrogenase, alanine dehydrogenase and reactive NAD^+ derivatives, (capable of removing ammonia by involving it into different synthetic reactions) is even more promising. It was shown that an extracorporeal reactor containing 1 g of hollow fibers with a immobilized multienzyme system decreases the ammonia content of the blood from 3 mg/ml to normal levels after 16 hours. In [664], a detoxifying reactor has been described for the removal of urea from blood, which consists of sequentially arranged columns with activated charcoal, immobilizing urease and a inorganic cation exchanger. The activated charcoal column achieves complete sorption of creatinine and partial sorption of urea; the column with immobilized urease almost quantitatively removes urea. The cation exchanger eliminates ammonia, formed in the urease-containing reactor.

Preparations of immobilized urease demonstrate the same rate of urea decomposition in citrate donor plasma as the native enzyme [665]. At the same time, the stability of immobilized urease (on diethylaminoethyl cellulose or a copolymer of cellulose with polyglycidyl methacrylate) was much higher than that of the native enzyme and permitted the repeated use of immobilized enzymes. The removal of urea from the blood by using an extracorporeal device with urease immobilized into foamed polytetrafluoroethyle membranes has been described [666]. A converter with urease immobilized on hemocompatible macroporous copolymers of styrene and devinylbenzene, has been suggested [667].

Data on urease immobilization in gelatine [668] and on carbon fibers [669] are also of particular interest, because both types of enzyme-containing materials are suitable for the use in extracoporeal devices.

Immobilization has also been used to increase the therapeutic efficiency of urate oxidase (uricase). This enzyme can play an important role in the last step of purine base metabolism. Sometimes, the process of uric acid elimination from the body can be disturbed and its accumulation causes different pathological reactions, including the development of gout. In man and some other mammals, uric acid is the final product of purine metabolism. In other mammalian species, uric acid is converted into allantoin by the action of liver urate oxidase. In man suffering from gout and other inherited diseases, sodium urate accumulates in the body, especially in the blood and cartilage. Furthermore, uric acid forms kidney stones. Inhibitors of xantine oxidase (the enzyme which converts xantine into uric acid) which are used for the treatment of these diseases are not always effective enough; besides, they do not act on the uric acid which has already been formed. Attempts were made to influence purine base metabolism by breaking down uric acid with urate oxidase, which is a Cu-containing enzyme (m.w.n. 420 000) and can be purified from mammalian liver or obtained by microbial synthesis.

Extracorporeal shunts with uricase immobilized on glass beads sharply decreases the uric acid level in blood and urine of experimental dogs with a simultaneous increase in the urinary content of harmless allantoin [670].

A variety of extracorporeal devices with immobilized enzymes has been suggested for the treatment of liver insufficiency and coma, caused by the accumulation of toxic metabolites like ammonia, mercaptanes, phenols, phenylalanine and free fatty acids in the blood [671].

The widely used blood purification by passing it through activated charcoal gives definite clinical results, but is usually accompanied with different side reactions because of non-specific sorption of many blood proteins and undesirable effects on blood cells. An attempt to solve this problem by coating activated charcoal particles with albumin or semipermeable polymeric membranes was also unsuccessful. The authors of [671] describe the use of reactors with immobilized glucuronyl transferase for the removal of glucuronic acid (in rabbit experiments); with immobilized phosphoadenosine-S-phosphosulfate transferase for the removal of sulfates; with immobilized acylthiokinase/transacylase for the coupling of glycine with organic acids; with immobilized glutathion-S-aryl (or acyl) transferases for the formation of mercaptopurinic acids; with immobilized S-methyl transferases for the methylation of mercaptones and with immobilized oxidases. A multifunctional device could solve completely the task of extracorporeal detoxification of the body from harmful metabolites.

According to the authors of [671], the immobilization of cytochrome P-450 and related detoxifying systems is of special interest, because these systems (due to their broad specificity) protect effectively the body from different endogenous and exogenous toxic compounds.

To remove the excess of bilirubin from the blood in the treatment of endogenous jaundice an extracorporeal device with immobilized bilirubin oxidase from Myrothecium verrucaria was proposed [672]. The reactor contained the enzyme immobilized on BrCN-activated Sepharose. The specific activity of the immobilized enzyme was 20 units/g gel; its thermostability was 5 times higher than that of the native enzymes. The column reactor with 15 ml of the enzyme-containing gel can oxidize 60% of abnormally elevated bilirubin in the cycle in an experimental media. In order to minimize the undesirable effect of the procedure on blood components, an extracorporeal reactor based on chemically cross-linked fibrin has also been studied [673].

In thrombolytic therapy the important problem is the fast removal of heparin from the blood flow after completion of the treatment course or in the case of heparin overdosing. To solve this problem, the authors of [674–676] suggested that an extracorporeal reactor with immobilized heparinase should be used (see also Figs. 56, 57). Heparinase, the enzyme specifically catalyzing heparin degradation into small saccharide molecules, had been immobilized on Sepharose 4B and the extracorporeal shunt containing the immobilized enzyme was incorporated into the circulation of experimental animals. The authors demonstrated the high efficiency of the system in the degradation of free heparin. The shunt had minimal effect on normal blood composition.

A system which simultaneously contains the whole set of enzymes responsible for the hepatic metabolic chains is expected to act closest to the physiological conditions. Thus, an "artificial liver" has been described [677], based on transformation of ammonia into urea by enzymes from the ornithin synthesis cycle together with carbamoyl phosphate synthetase I and inorganic phosphatase, immobilized on fibrin membranes. The system can effectively function with human plasma. Urea is supposed to degrade further to glyoxylic acid by another multienzyme system, consisting of urease, catalase, allantoinase and allantoicase [677, 678].

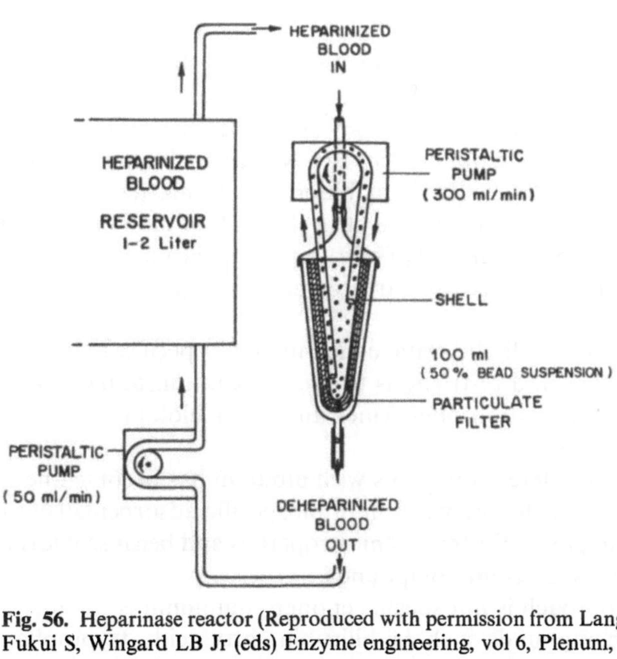

Fig. 56. Heparinase reactor (Reproduced with permission from Langer R et al. (1982). In: Chibata I, Fukui S, Wingard LB Jr (eds) Enzyme engineering, vol 6, Plenum, New York, p 438)

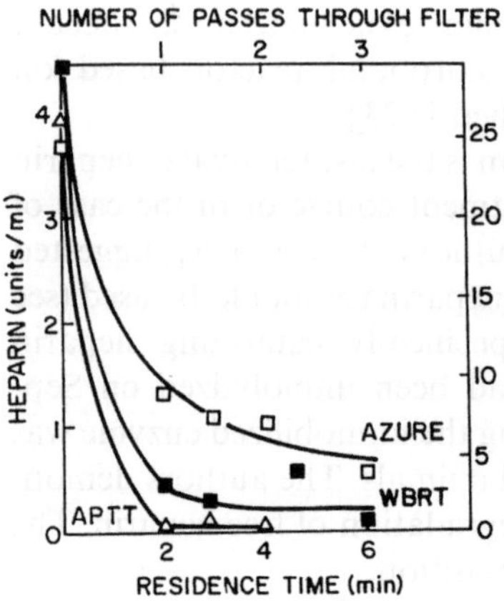

Fig. 57. Heparin levels in dog blood after administration of 4500 u. of heparin. Blood samples were analyzed by 3 assays: APTT, whole blood recalcification time and Azure A (Reproduced with permission from Langer R et al. (1982). In: Chibata I, Fukui S, Wingard LB Jr (eds) Enzyme engineering, vol 6, Plenum, New York, p 439)

Enzymes of the urea cycle are of special interest from the point of view of their coimmobilizaiton with other enzymes to obtain "physiological" extracorporeal detoxifying devices. In [679–681], different methods for their immobilization are described together with possible applications of immobilized enzymes in extracorporeal reactors. The authors of [679] suggested the immobilization of all four enzymes of the urea cycle—carbamoyl transferase, arginine succinate synthetase, arginine succinate lyase and arginase, together with inorganic pyrophosphatase. Immobilization was performed on 6-aminohexyl Sepharose by glutaraldehyde coupling. This system is capable of utilizing arginine with the formation of fumarate—the final product of the whole cycle. The same enzymes can be non-covalently incorporated into a fibrin matrix [680]; their practical use for detoxification of ammonia has been discussed [681].

Many studies utilize affinity chromatography principles in extracorporeal devices. For example, albumin immobilized on BrCN-activated Sepharose has been used for the removal of bilirubin from blood or plasma (especially in newborns). Experiments on rats with jaundice have shown that blood perfusion through this gel leads to complete removal of bilirubin from plasma in one hour. Similar data were confirmed by successful treatment of jaundice in new-born monkeys [682, 683].

We will not describe here all the data concerning biospecific polymeric adsorbents; a detailed review and restrictions to the preparation of biospecific adsorbents, based on immobilization of enzymes and other biologically active proteins can be found in [684].

Recently, the coating of different materials with albumin has become one of the most popular methods for the preparation of biospecific adsorbents [684], mainly because albumin imparts athrombogenic properties and because it forms stable complexes with many different compounds.

Unfortunately, this approach is not specific enough—albumin is capable of binding many different compounds such as: bilirubin, fatty acids, tryptophan, metal ions and many drugs (e.g., aspirin and some antibiotics). Attempts to use synthetic polymers to mimic albumin have been unsuccessful, because the specificity of the systems obtained was even worse than that of native albumin; besides, the binding constants were also lower.

Many experiments have been performed with adsorbents containing immobilized proteolytic enzymes (e.g., trypsin) or their inhibitors [684]. These adsorbents can be used in the treatment of acute pancreatitis by hemosorption. In a normal pancreas, the simultaneous release of enzymes and inhibitors protects the gland from autolysis. In acute pancreatitis, proteolytic enzymes are liberated and accumulate in the pancreas, which leads to the selfdigestion of pancreatic tissue. Products of this selfdigestion are released into the circulation and cause a dangerous intoxication. The most pronounced increase in activity was found for trypsin (10 to 40 fold); an obvious mode of therapy is the removal of enzyme excess by passing blood plasma through devices with biospecific adsorbents, containing immobilized trypsin inhibitors. This approach has been studied experimentally. The use of hemodialyzers containing the immobilized pancreatic trypsin inhibitor, ovomucoid and several other highly specific trypsin inhibitors lead to a pronounced decrease of trypsin activity in the perfusate, simplified the therapeutic procedure and made it less expensive with a sharp decrease in mortality [685, 686].

A system which occupies an intermediate position between traditional extracorporeal devices and artificial cells also exists. This system utilizes microencapsulated enzymes as fillers for extracorporeal shunts. Extracorporeal perfusion through microencapsulated enzymes can be applied for the transformation and removal of toxic metabolites from biological fluids. It is used in artificial kidneys or livers and in the treatment of acute intoxications. The system has some evident advantages: the microcapsules do not leave the reactor, which excludes polymer accumulation in the organism, they can be easily renewed or replaced and reactors with microcapsules are very compact, in comparison with large dialyzers (which usually require large quantities of dialyzing solutions—up to 100–200 liters.). Besides, the thickness of microcapsules membrane is less than that of dialyzing membrane by 100-fold, which increases the diffusion rate of toxic substances 200 to 1000-fold. Ultrathin microcapsules can be used at perfusion rates as high as 200 ml/min.

All the methods clearing urea from the circulation are based on reactors with microcapsules, containing urease and ion-exchangers or multienzyme systems [191, 193, 687, 688]. Microcapsules, containing urease and an ion-exchanger, convert urea into ammonia and then remove the latter. It was also suggested to use positively charged microcapsules which repulse cations and thus prevent NH_4^+ transport from the capsules. In a microencapsulated multienzyme system, containing urease, glutamate dehydrogenase and glucose-6-phosphate dehydrogenase, urease converts urea into ammonia, which interacts with α-ketoglutarate in the presence of NADPH, yielding glutamate. The necessary optimal ratio between NADPH and $NADP^+$ is maintained by glucose-6-phosphate dehydrogenase [203, 689].

Table 19. Enzymes immobilized extracorporeally that demonstrate clinical feasibility (a partial list). (From Klein M.E. and Langer R., TIBTECH, July 1986, 179, with permission)

Enzyme	Disease	Reactor	Results
β-Galactosidase	Fabry's disease	Sepharose 48	Not active in blood though active in other solutions
Arginase	Familial hyperargininemia	External surface of Amicon (R) fibers in a chamber; substrate diffuses in a rabbit model	Urea production increased until enzyme inhibited by product. Reactor could be regenerated to 85% of initial activity by washing after 24 h blood exposure
Asparaginase	Cancer	External surface of fibers in Cordis-Dow^R hollow fiber hemodialyser	Reduced asparagine in dog from 50 um to <2 um in 30 min and for up to 4 h. Used without clinical problem in a human known to have anaphylactoid response to asparaginase
		Methacrylate plates	Removed asparagine in two patients with melanoma and no antibodies detected with Ouchterlony double-diffusion assay
Bilirubin oxidase	Liver failure	Sepharose in a packed bed reactor	Decreased serum bilirubin in Gunn rats by 50% in 30 min
Carboxypeptidase G1	Tumors treated with methotrexate, e.g. osteogenic sarcoma	Polysulfone hollow fibers	Reduced methotrexate levels in non-blood solution from 100 um to (2 um in 30 min. No enzyme leakage. Stable at 37 °C for 4 h; at 40 °C for 9 months
Heparinase	Removing heparin during extracorporeal circulation	Sepharose 4B in a fluidized bed reactor	All anticoagulant action removed in 2 min from dogs connected to reactor with blood flow 50 ml/min. No antibodies detected by Ouchterlony double diffusion assay
Liver microsomal enzymes	Liver failure	Large agarose beads in packed bed reactor	Glucuronidation of phenol in rabbits
Phenylalanine ammonia lyase	Phenylketonuria	Amicon^R hollow fibers	Reduced phenylalanine in dogs and monkeys with experimentally induced phenylketonuria. No adverse hematological, immunological, or physiological effects during 2 h treatment
UDP-Glycuronyl transferase	Liver failure	Agarose beads	80% of enzyme activity recovered; increased stability; able to conjugate p-nitrophenol and 1-naphthol
Urea cycle enzymes	Liver failure	Blood dialysed through hollow fibers, cell and protein free fluid to packed bed Sepharose 4B reactor	

A detailed review on microencapsulated urease in extracorporeal shunts is presented in [690, 691], where the application of microcapsules with urease in an "artificial kidney" apparatus was considered. It was shown that the diffusion rate of low molecular weight metabolites in the device with microcapsules is three orders of magnitude higher than in a membrane reactor with a comparable membrane surface area. A routine "artificial kidney" apparatus has a surface area of the dialysis membrane of approx. 1 m^2. Ten ml of microencapsulated urease suspension with a microcapsule diameter of 20 μm each have a surface area of $\sim 2.5\, m^2$. In dog experiments, the urea lend in blood was decreased by 50% after 1.5 hour of blood recirculation through a shunt with microencapsulated urease.

Other examples of the use of microencapsulated enzymes in extracorporeal shunts exist. A device with microencapsulated catalase was found to be effective in the replacement therapy of acatalasemic mice [191, 193]. The extracorporeal system, consisting of multiple capillaries and microcapsules with immobilized asparaginase, effectively removed asparagine from the blood [193]. Heparinization of the system and aditional stabilization of the entraped enzyme prolongs the activity of the hemoperfusion system. Circulating DNA can be degraded in extracorporeal reactors/with nylon microcapsules containing immobilized nuclease [690].

Some results discussed above are summarized in Table 19.

5.2 *Dressing and Drainage Materials Containing Immobilized Enzymes; Films, Ointments and Powders with Immobilized Enzymes in the Treatment of Wounds, Ulcers and Burns*

Immobilized enzymes can be also used for the manufacturing of improved dressings, sutures, drainages and similar materials. The enzymes (usually proteolytic), have been immobilized on the surface or inside appropriate textiles. In the treatment of external injuries, such materials provide antibacterial activity, lysis of necrotic tissues and accelerated healing.

Fibrous materials, containing immobilized proteolytic enzymes (trypsin, terrilitin, protosubtilin, chymotrypsin) are of particular interest for clinical application. Such materials were obtained by chemical coupling of proteolytic enzymes to fibrous materials [694, 695] or by enzyme incorporation into fibers in the process of their formation [696].

The following materials were used for the covalent coupling of proteolytic enzymes: cellulose fibers oxidized with iodic acid (so-called dialdehyde cellulose); cellulose, polyvinyl alcohol and collagen fibers treated with bromal; polyvinyl alcohol fiber treated with maleic dialdehyde; hydrolyzed polyamide fibers treated with glutaric dialdehyde [694–696].

Water-soluble polymeric derivatives of α-chymotrypsin, containing 60% of enzyme, were obtained by enzyme coupling to carboxymethyl dextran with a m.w. of 60 000 and 100 000 via 2-amino-4-chlor-s-triasine residues; the activity of the bound enzyme was 90% of the initial value [697]. Polymeric derivatives of α-chymotrypsin, containing 9% wt. of the chemically bound enzyme (with an activity of about 20% of the initial value) were obtained with blue dextran by the same method. When coupled to carboxymethyl dextran esters, α-chymotrypsin increased its thermostability. When α-chymostrypsin had been incorporated into a film made of cellulose triacetate (CTA) during the process of film formation, the enzyme was almost completely desorbed in 1 M NaCl solution. At the same time, only a small amount of enzyme was desorbed from the films when α-chymotrypsin was first coupled to carboxymethyl dextran ester (100 000) or to blue dextran and then incorporated into the CTA film. In this case, the enzyme thermostability was also increased.

Numerous methods of terrilitin immobilization on cellulose and polyacrylamide fibrous materials have been developed. Ionic and/or covalent immobilization were performed on modified cellulose fibers, containing aldehyde and carboxyl groups. Glutaric dialdehyde, a bifunctional cross-linking reagent, was used for terrilitin fixation on non-modified cellulose and polyacrylamide materials. The materials obtained did not loose their proteolytic activity after sterilization by γ-radiation and storage. Fibrous materials containing immobilized terrilitin proved to be effective for the treatment of purulent wounds [694]. The same is true for trypsin.

Currently, cellulose and polyacrylamide fibers containing covalently bound trypsin and polyvinyl alcohol fibers containing electrostatically bound trypsin proved to be effective in reducing the time of cleaning and healing of purulent wounds and for the treatment of parodontosis [694–696].

It was established in [698] that trypsin, incorporated into the structure of ftorlon filaments in the process of filament formation, can diffuse from the fiber into a model liquid medium. Suture materials made of such filaments prevent development of experimentally induced infections in animals, accelerate wound cleaning of microorganisms, and reduce the time of wound healing. Modified gauze, containing covalently bound hyaluronidase, proved to be effective for the treatment of patients with acute purulent mastitis.

Enzyme immobilization by its mechanical incorporation into the fiber structure in the process of fiber formation has been demonstrated long ago. In [699], β-galactosidase was incorporated into regenerated cellulose fibers. The immobilized enzyme exhibited increased stability.

A general method for the fixation of proteolytic enzymes retaining their activity on dressing materials was proposed: to fix enzymes on medical gauze, synthetic fabric (50% viscose and 50% lavsane) and non-woven bandages using polyvinyl alcohol [700]. For this purpose, the required textile was soaked with the enzyme in a poyvinyl alcohol solution. Then the fabric was dried and sterilized by 2.5 Mrad, γ-radiation. Table 20 demonstrates that the method suggested provides sufficient binding of proteins with the dressing material, retaining a high enzymatic activity even after sterilization. Such materials significantly accelerate the healing of experimental burn wounds in rabbits.

Table 20. Immobilization of proteolytic enzymes on dressing materials [700]

Enzyme	Dressing material	Active enzyme, mg/g carrier	Immobilized enzyme activity, units/g carrier	Activity yield upon immobilization, %	Activity preservation upon sterilization, % from initial
Trypsin	gauze bandage	7.1	43.3	59.0.	65.6
Papain	—"—"—"	12.2	17.1	31.7	57.7
Bacterial protease	—"—"—"	19.9	71.6	70.4	
—"—"—"	—"—"—"	24.6	88.6	63.4	82.1
—"—"—"	—"—"—"	40.6	146.2	80.3	
—"—"—"	Synthetic cloth	18.0	65.0	71.4	93.6
—"—"—"	—"—"—"	43.7	157.4	69.4	80.9
—"—"—"	non-woven bandage	9.7	34.9	52.8	—

Analogous experiments have been performed with materials of various chemical composition. Proteinases, immobilized on dressing materials made of mixtures of cellulose or/and polycapramide fibers with non-woven carbonic fabric or with collagen, are the most promising in wound treatment [694].

It has been already demonstrated that surgical suture material, made of trypsin-containing fibers, markedly decreases microbial contamination of both clean and infected wounds and thus accelerates the healing process. Films made of cellulose acetate and other water-insoluble polymers with incorporated proteolytic enzymes may be used as drainage materials. Such films exhibit pronounced and prolonged drainage ability, as well as necrolytic activity, due to the gradual release of small amounts of active proteases. Films, coating wounds after one application can be used for a long period of time, which excludes the degradation of granulating tissue formed during healing. The use of such drainage materials accelerates wound cleaning of purulent and necrotic masses 3 to 4 times when compared with gauze or rubber drainages. Modified films which contain incorporated proteolytic enzymes and have the additional ability to dissolve slowly in water provide a long lasting enzyme depot and reduce significantly the period of purulent wound cleaning when compared with the effect of native enzymes. The result achieved is explained by the enhancement of reparative processes in wounds due to faster necrolysis by the action of stabilized proteases, immobilized on a polymeric carrier.

The first results with the experimental application of the above materials seem to be very promising. It has been demonstrated that the incorporation of 1–5% of proteolytic enzymes into dressing textiles shortens the period of wound healing by 4 to 6 days, reduces the cost of expensive enzymes 6 to 7 times (as compared with the native form), and reduces the period of necrotization and healing 1.5 to 3.0 times [701]. It has also been demonstrated that when various suture materials are used in experimental surgery, the least tissue reaction was observed after use of ftorlon, containing immobilized trypsin; furthermore, the least number of suppurations (about 14%) has been observed with this material (40%, when silk thread is used). The formation of inflammatory granulomas around silk threads in infected wounds can be observed on the 7 to 10th day in skin-muscle sutures, whereas in the case of trypsin-containing fibers, this process starts on the 5th day [702].

A clinical-experimental study on the possible use of proteases (trypsin and terrilitin) immobilized on textile materials in the treatment of purulent inflammatory deseases of soft tissues has been carried out. 630 patients have been studied; 420 hospitalized patients and 210 patients in an out-patients' clinic [703] have been treated with positive results.

Cotton wool, containing trypsin and other non-enzymatic therapeutics (such as furadonin, gumizol, hexamethason) has been clinically tested in stomatology for the treatment of regional parodontosis (inflammatory form) and mucuous membranes of the oral cavity, such as ulcerative-necrotic stomatitis, multiple exsudative erythema, various non-specific ulcers of the oral cavity, and aphthous stomatitis [704]. The immobilized preparations significantly reduced the length of the treatment period as compared with the native form of trypsin and other drugs. In patients with parodontosis, the purulent discharge from the dental pockets decreased by the time of the 2 or 3rd visit; it completely disappeared by the 4 to 5th visit. Cleaning of ulcer surfaces and active regeneration of the mucuous membranes were observed in patients with mucuous pathology of the oral cavity together with an analgesic effect.

A suspension of protease, immobilized on biocompatible microgranulated carriers, when applied on purulent wounds, is capable to hydrolyse denatured structures, tissue and cell patches within 10 days without loss of activity from the wound secretions [705]. Similar experiments on animals demonstrated that stimulation of granulation and epithelization processes by proteolytic enzymes immobilized on solid powder carriers reduced the time of wound healing two fold. Such results have been already clinically documented [705].

The biocompatibility of polyethylene glycol was used for manufacturing of powder-like materials containing immobilized enzymes which aid wound healing. Particles of polyethyleneglycol gel, size 100 to 250 μm and containing proteolytic enzymes (chymotrypsin, trypsin and fibrinolysin) were added to powders and ointments used for acceleration of wound healing [706]. In some cases they were additionally modified with poly-β-alanine.

The most interesting theoretical, experimental and clinical results have been reported by the authors of [705]. They have designed, introduced and studied in detail in veterinary and clinical practice a new drug, Profesim, which is a complex of alkaline and neutral proteases from Bacillus subtilis, covalently bound with water-insoluble aminoethyl cellulose. It was applied for treatment of purulent processes in soft tissues, as well as lungs, pleura and other localizations. The ability for long-term hydrolysis of proteins (over 10 days) with unchanged proteolytic enzyme activity is considered to be an important property of Profesim.

It has been demonstrated that Profesim favours intensive cleaning of purulent wounds and open abscesses from non-viable tissue, pus and blood clots.

Profesim does not activate the proteolytic activity of granulating tissue. It stimulates reparative processes presumably by removal of non-viable fragments. It was found to dramatically increase the efficiency of a complex treatment of pleural empysema in combination with antibiotics, flushing the cavity with antiseptic solutions, drainage, etc. Profesim was administered every day intrapleuraly in a dose of 2–8 g with drainage. This lead to accelerated evacuation of exsudate and fibrin clots, improvement of the patients' condition, his blood parameters and normalization of temperature. The enzyme therapy caused no complications. On the whole, Profesim shortened the treatment of pleural disease by half.

Clinical trials have demonstrated that a single endobronchial administration of Profesim to the postoperative patients with lung pathology intensified the evacuation of sputum (which became less viscuous) within three hours. Combination of the enzyme preparation with antimicrobial agents proved to be an important part of the therapy.

The authors performed comparative histological and histochemical studies on the dynamics of standard skin-muscular wound healing in experiments on male Wistar rats and on clinical material (post surgical festering and nonhealing wounds) after treatment with traditional methods and with Profesim.

The cleaning of purulent wounds of necrotic tissues in rats was completed after 5 to 6 days by Profesim; in control animals, this effect was observed on the 10th day. Predominance of fibroblasts in granulating tissue, onset of secretory activity, reduction of acute inflammation, de novo formation of collagen fibrilles in the upper layer of granulating tissue of the wound, formation of a cellular pauch at the bottom and on the edges of the wound and effective epithelization were observed 3–4 days earlier after treatment with Profesim.

1058 patients with purulent post-operational wounds, purulent mastitis, phlegmonas, and pulmonary and pleural purulent processes were subjected to protracted treatment with proteolytic enzymes. It was found that Profesim exhibited a prolonged proteolytic effect and caused no local and general reactions of the body. Surgical necrotomy in combination with Profesim therapy shortened the recovery period.

Profesim effectively hydrolyses denaturated proteins; purlent-necrotic wounds were cleaned on average 1.5–2 times faster than with soluble proteolytic enzymes. A single administration of the preparation can continuously hydrolyse regions of injury for 5–6 days.

A water-soluble preparation, "Immozin", with a proteolytic activity of up to 100 PU/ml, with casein as substrate, and high thermostability in the presence of substrate (37 °–80 °C) was obtained by immobilization of bacterial proteases on 1,4-polyalkylenoxide. Immoxin retained its activity at pH from 6.0 to 11.0.

Both preparations have been successfully used for "chemical" necrotomy of wounds. The enzyme activity of both preparations was maintained when combined with routinely used antibiotics. This permits the treatment of purulent

Table 21. Some examples of immobilized therapeutic enzymes. (From Torchilin V.P., Adv. Drug Del. Rev., 1, 41 1987, with permission

Enzymes (EC)	Intended Use	Carrier	Comments
Acetylhexoseaminidase	Tay-Sachs' disease	soluble activated polyvinyl pyrrolidone	animal experiments
δ-aminolevulinate dehydratase	porphyriasis	liposome	
α-Amylase	burns, abscesses, digestive insufficiency	liposome	amoeba
		soluble cyanogen bromide activated dextran	rat
		glutenin	
		insoluble cross-linked polymers containing methacrylic acid residues	
Amyloglucosidase	type II glucogenosis storage diseases	liposome	man, rat
Arginase	oncology	soluble cyanogen bromide activated dextran	mice
Asparaginase	oncology, blood treatment	insulin-globulin albumin	mice, rat
		insoluble polyacrylamide and polydextran microspheres	mice
		collagen	
		liposome	mice
		microcapsules	rat, mice
		red blood cells ghosts	
		modified polyacrylamide microgranules	
Carboxypeptidase	oncology	soluble cyanogen bromide activated dextran	mice
Catalase	burns, abscesses, wound healing, acatalasemia	soluble cyanogen bromide activated dextran	rat, mice
		polyethyleneglycol	
		microcapsule	
		liposome	rat fibroblasts
		microgranules made of polyvinylpyrrolidone, polyvinylalcohol, or their copolymers	
Cholesteroloxidase	atherosclerotic deposits	soluble polymer	model targeted systems
Collagenase	rheumatism, atherosclerotic deposits, adjuvant arthritis	dextran	rat
Cytochrome oxidase	oncology	liposome	
Desoxyribonuclease	inflammation	insoluble polymer	
β-D-fructofuranosidase	storage diseases	liposome	rat

α-galactosidase	Fabry's disease	albumin	
β-galactosidase	Fabry's disease	intermolecular cross-linking	
		red blood cell ghosts	
		liposome	fibroblasts
		microcapsule	mice
α-glucosidase	Pomp's disease	soluble polymer; low density lipoprotein complex albumin	
β-glucosidase	Gaucher's disease	liposome	man
	Krabbe's disease, gangliosidosis		
		red blood cell ghosts	
Glucose oxidase	oncology	liposome	
		collagen	
		adsorption on rat platelets; human red blood cells	
β-glucuronidase	rheumatism, oncology, adjuvant arthritis	soluble polymer	
		liposome	mice
		red blood cell ghosts	mice monkey
Glutaminase	liver diseases, oncology	asialoorozomucoide	
Glutaminase-L-asparaginase	liver diseases, oncology	soluble polyethyleneglycol	mice lymphosarcoma
		concanavalin A-Sepharose	
Hexoseaminidase A	G_{M2} gangliosidosis, Tay-Sach's disease	liposome	
Kallikrein	inflammation	soluble activated polyvinylpirrolidone	dog
Lactase	inherited lactase deficiency	microcapsule	
	liver diseases	covalent binding to polymers	
Lysozyme	inflammation, liver diseases	collagen	animal experiments
		activated dextran	
		asialofetuin	
		liposome	
		microcapsules made of polyvinylpirrolidone, polyvinyl alcohol or their copolymer	
Mannase	rheumatism, adjuvant arthritis	soluble dextran	rat
α-mannosidase	glycoproteinosis	liposome	
Neuraminidase	sialidosis	liposome	rat
Nitrate reductase	oncology	microcapsule	mice lymphosarcoma
Peroxidase	peroxidase insufficiency	liposome	rat
		adsorption on rat platelets and human red blood cells	

(*Continued*)

Table 21. (*Continued*)

Enzymes (EC)	Intended Use	Carrier	Comments
Phenylalanine ammonia lyase	oncology	soluble polymer	mice
		microcapsule from cellulose nitrate	mice
Plasmin	thrombolytic therapy	soluble dextran	animal experiments
		oxidized Sephadex	dog
Ribonuclease	inflammation, oncology	intermolecular cross-linking	rat
		insoluble polymer	
Streptokinase	thrombolytic therapy, opthalmological pathology	insoluble polymers	
		oxidized Sephadex	
		soluble oxidized dextran	dog man
Superoxide dismutase	toxic shock	soluble polymer	
		liposome	rat
Trypsin	burns, wound healing, blood kallikrein activation	soluble polyethyleneglycol	animal experiments
		insoluble polymer, dressing material	
Thrombin	cardio-vascular diseases	insoluble polymer	
Tissue plasminogen activator	thrombolytic therapy	soluble polymers, antibodies	lysis of fibrin clot in vitro, animal experiments
Urate-oxidase	detoxication	polyacrylamide, polyhydroxy-methylmetacrylate, polyamide and polyisocyanate gels	
Urease	blood detoxication, diabetes	soluble polymer, insulin, microcapsule, liposome, insoluble polymer	animal experiments
		insulin	
		microcapsule	
		liposome	
		insoluble polymer	
Uricase	oncology, blood detoxication	soluble polymer	mice
		collagen	
		microcapsule	
Urokinase	thrombolytic therapy	soluble polymer	
	athrombogenic materials for prosthetic appliance	soluble cyanogen bromide activated dextran	lysis of fibrin clot in vitro
		oxidized Sephadex	

Table 22. Enzymes immobilized intracorporeally that demonstrate clinical feasibility (a partial list). (From Klein M.D. and Langer R., TIBTECH-July 1986, 179, with permission)

Enzyme	Disease	Carrier	Results
β-1,4-Glucosidase	Type II glycogenosis or Pompe's disease	Albumin	Increased resistance to heat denaturation and to proteolysis by trypsin
Arginase	Familial hyperargininemia	Polyethylene glycol	Active; less immunogenic than free enzyme; longer half-life in circulation; can be absorbed from peritoneal cavity
Asparaginase	Cancer	Polycrylamide and polyacryldextran microcapsules i.m. and i.p.	Asparagine rapidly removed from circulation. Survival with complete tumor repression in up to 10 of 12 mice with lymphoma
		Artificial cells	Suppressed lymphosarcoma in rats. No antibodies detected by Ouchterlony double diffusion assay
		Red blood cell ghosts	Active; half-life increased to 8 days
		Polyethylene glycol	Active; less immunogenic than free enzyme; longer half-life in circulation; can be absorbed from peritoneal cavity
		Liposomes	Mice sensitized to asparaginase did not show evidence of adverse immunological reactions when give i.v.
β-Glucuronidase	Mucopolysaccharidosis VII	Red blood cell ghosts	In a mouse model enzyme was active, not immunogenic, and longer-lived than soluble form
Catalase	Acatalasemia; inflammatory diseases mediated by oxygen free radicals	Collodion microcapsules	Protected acatalasemic mice from a lethal challenge with sodium perborate even in those with antibodies to catalase
		Polyethylene glycol	Active; less immunogenic than free enzyme; longer half-life in circulation; can be absorbed from peritoneal cavity

(*Continued*)

Table 22. (*Continued*)

Enzyme	Disease	Carrier	Results
Fibrinolysin	Thromboembolic occlusive vascular disease	Sephadex	Required about 99% less enzyme to clear clotted dog femoral artery than for soluble form
Glutaminase	Cancer	Polyethylene glycol	Active; less immunogenic than free enzyme; longer half-life in circulation; can be absorbed from peritoneal cavity
Phenylalanine ammonia lyase	Phenylketonuria	Polyethylene glycol	Active; less immunogenic than free enzyme; longer half-life in circulation; can be absorbed from peritoneal cavity
Profezyme (bacterial protease)	Cystitis	Aminoethylcellulose	Ameliorated symptoms and gross pathology of chronic cystitis
Soil enzymes	Renal failure	Microcapsules	Metabolism of nitrogenous wastes and removal of potassium
Superoxide dismutase	Inflammatory diseases mediated by oxygen free radicals	Polyethylene glycol	Active; less immunogenic than free enzyme; longer half-life in circulation; can be absorbed from peritoneal cavity
Tyrosinase	Liver failure	Collodion artificial cells	Tyrosine reduced to 50% and phenol lowered after 1 h perfusion
Uricase	Gout	Polyethylene glycol	Active; less immunogenic than free enzyme; longer half-life in circulation; can be absorbed from peritoneal cavity
Urokinase	Small diameter vascular graft	Fibrocollagenous tube	Locally activated fibrinolysin so that thrombosis less than in tubes without enzyme bound
	Thromboembolic occlusive vascular disease	Agarose beads	Enzyme activity varied with length of spacer interposed between enzyme and carrier

wounds by combination of traditional surgical and antiseptic methods with prolonged enzyme therapy.

The application of immobilized enzymes in therapy is reviewed in Tables 21 and 22; the first presents general data on the clinical application of immobilized enzymes, while the second illustrates in detail data on the most important preparations. We hope that in time more names will be included in such tables.

6 Immobilized Enzymes in Clinical Analysis

The enormous body of data on the analytical application of different systems of immobilized enzymes, is traditionally considered to be an independent field of enzyme engineering. Many books and reviews have been devoted to analytical methods based on immobilized enzymes. Here we shall only consider data on general principles. Various enzyme-coated devices (columns with enzyme immobilized on microparticles; enzymes immobilized on the inner surface of reactors or incorporated into hollow fibers or membranes) serve to perform hundreds of analyses with the same amount of enzyme in the same equipment. Many analytical systems based on immobilized enzymes can be automatized. In particular, an automatic analyzer by "Technicon" is widely in use for clinical analysis. In this analyzer, the duration of a particular analytical reaction is determined by the rate of reagent delivery into the system.

Among the substances that can be discovered or determined quantitatively by means of immobilized enzymes, many are of particular interest in medicine: low molecular weight metabolites (glucose, urea, uric acid, amino acids, lipids, cholesterol) various drugs and toxic compounds, antibodies, toxins and heavy metals. High specificity of the analytical procedure, good reproducibility and sensitivity are the main advantages of the devices based on immobilized enzymes.

The most important methods of analysis, based on certain reactions catalyzed by immobilized enzymes, are the following: the use of enzyme electrodes, immunoenzyme analysis, bioluminiscent analysis, analysis in flow reactors and some other more exotic methods.

6.1 Enzyme Electrodes

Enzyme electrodes have been developed and described at the very beginning of enzyme engineering and are electrochemical transducers of various types, coated with a layer of immobilized enzyme. Their action is based on the determination of the product of enzymatic reaction or on the registration of changes in substrate levels in contact with the electrode. After contact of the electrode with the analyzed solution, the enzymatic reaction is proceeding continuously. If the substrate or product of this reaction are electrochemically active, changes of the electrode potential indicate the amount of substance determined. The continuous

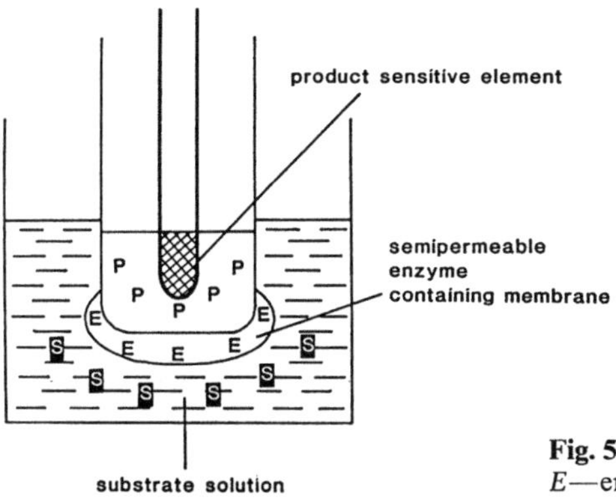

Fig. 58. The general scheme of an enzyme electrode. *E*—enzyme, *S*—substrate, *P*—product

determination of analyzed metabolites is one of the advantages of enzyme electrodes (scheme of electrode action, Fig. 58).

The enzyme electrode is not "enzymatic" in the full sense of the word: the enzyme is not the recording element. The function of the enzyme in this electrode is to change a local concentration of a certain product which is measured by some routine electrochemical sensor. Thus, the enzyme electrode unites two more or less independent aspects: a device with an immobilized enzyme and a transducer (sensor) which may even localized inside. Therefore, the construction of the electrode is of particular importance. Various enzymes are suitable for enzyme electrodes, and enzyme-containing elements of electrodes can be obtained by such methods as enzyme incorporation into polymeric gel, membrane or film, or immobilization in the space between the electrode and membrane. Sometimes, the enzyme solution is placed between two semipermeable membranes: one separates the enzyme from the sensor element, the other separates the enzyme from the solution analyzed. Enzyme electrodes exhibit good sensitivity (from 10^{-2} up to 10^{-9} M), reproducibility and electrical characteristics. At the same time, it should be noted that they are not widely used due to the complexity of their manufacturing.

Nevertheless, enzyme electrodes proved to be promising for the determination of certain metabolites and pharmacological agents [707]. In particular, in this very informative review a five-step mechanism of enzyme electrode action is formulated: 1) the substrate transport to the electrode surface; 2) the substrate diffusion, via membrane or gel, to the active site of the immobilized enzyme; 3) the reaction of substrate conversion; 4) the diffusion of the product to be determined (formed as a result of enzyme-catalyzed reaction) to the electrode surface 5) the quantitative determination of the product on the electrode surface.

Currently, more than 100 various enzyme electrodes have been described [708]. Table 23, based on [707] presents properties of some typical enzyme electrodes. More details on the properties of some of these electrodes are described in Table 24, such as enzyme immobilization methods and the type of sensors used. Table 25 enumerates the most frequently used electrode sensors.

Table 23. Typical electrodes and their characteristics [701]

Type	Enzyme	Sensor	Immobilization[a]	Stability	Response time	Amount of enzyme (U)	Range (mol/l)[b]
1. Urea	Urease (EC 3.5.1.5)	Cation	Physical	3 weeks	30–60 sec	25	10^{-2}–5×10^{-5}
		Cation	Physical	2 weeks	1–2 min	75	10^{-2}–10^{-4}
		Cation	Chemical	> 4 months	1–2 min	10	10^{-2}–10^{-4}
		pH	Physical	3 weeks	5–10 min	100	5×10^{-2}–5×10^{-5}
		Gas (NH_3)	Chemical	4 months	2–4 min	10	5×10^{-2}–5×10^{-5}
		Gas (NH_3)	Chemical	20 days	1–4 min	0.5	10^{-2}–10^{-4}
		Gas (CO_2)	Physical	3 weeks	1–2 min	25	10^{-2}–10^{-4}
2. Glucose	Glucose oxidase	pH	Soluble	1 week	5–10 min	100	10^{-1}–10^{-3}
		Pt(H_2O_2)	Physical	6 months	12 sec kinetic[c]	10	2×10^{-2}–10^{-4}
	(EC 1.1.3.4)	Pt(H_2O_2)	Chemical	> 14 months	1 min steady-state	10	2×10^{-2}–10^{-2}
		Pt(H_2O_2)	Soluble		1–2 min	10	10^{-2}–10^{-4}
		Pt(quinone)	Soluble	< 1 week[d]	3–10 min	10	2×10^{-2}–10^{-2}
		Pt(O_2)	Chemical	> 4 months	1 min	10	10^{-1}–10^{-5}
		I^-	Chemical	> 1 month	2–8 min	10	10^{-3}–10^{-4}
		Gas (O_2)	Physical	3 weeks	2–5 min	20	10^{-2}–10^{-4}
	Glucose oxidase (EC 1.1.2.4) and Catalase (EC 1.11.1.6)	Gas (O_2)	Chemical	> 3 weeks	2–5 min	10	2×10^{-2}–10^{-4}
3. L-Amino acids (general)[e]	L-amino acid oxidase (EC 1.4.3.2)	Pt(H_2O_2)	Chemical	4–6 months	12 sec kinetic[c]	10	10^{-3}–10^{-5}
		Gas (O_2)	Chemical		2 min	10	10^{-2}–10^{-4}
		Pt(O_2)	Chemical	> 4 months	1 min	10	10^{-2}–10^{-2}
		Cation	Physical	2 weeks	1–2 min	10	10^{-2}–10^{-4}
		NH_4^+	Chemical	> 1 month	1–3 min	10	10^{-2}–10^{-4}
		I^-	Chemical	> 1 month	1–3 min	10	10^{-3}–10^{-4}
L-Tyrosine	L-Tyrosine decarboxylase (EC 1.1.25)	Gas (CO_2)	Physical	3 weeks	1–2 min	25	10^{-1}–10^{-4}
L-Glutamine	Glutaminase (EC 3.5.1.2)	Cation	Soluble	2 days[d]	1 min	50	10^{-1}–10^{-4}

L-Glutamic acid	Glutamate dehydrogenase (EC 1.4.1.3)	Cation	Soluble	2 days[d]	1 min	50	10^{-1}–10^{-4}
L-Asparagine	Asparaginase (EC 3.5.1.1)	Cation	Physical	1 month	1 min	50	10^{-2}–5×10^{-5}
4. D-Amino acid[f] (general)	D-amino acid oxidase (EC 1.4.3.5)	Cation	Physical	1 month	1 min	50	10^{-2}–5×10^{-5}
5. Lactic acid	Lactate dehydrogenase (EC 1.1.1.27)	Pt([Fe(CN^{4-}]	Soluble	< 1 week	3–10 min	2	2×10^{-3}–10^{-4}
6. Succinic acid	Succinate dehydrogenase (EC 1.3.99.1)	Pt(O_2)	Physical	1 week	1 min	10	10^{-2}–10^{-4}
7. Acetic, formic acids	Alcohol oxidase (EC 1.1.3.13)	Pt(O_2)	Chemical	> 4 months	30 sec	10	10^{-1}–10^{-4}
8. Alcohols	Alcohol oxidase (EC 1.1.3.13)	Pt(H_2O_2) Pt(H_2O_2) Pt(O_2)	Soluble Soluble Chemical	1 week 1 day[d] > 4 months	12 sec kinetic[c] 1 min 30 sec	10 1 10	0.5–100 mg% 0.5–50 mg% 0.5–100 mg%
9. Pencillin	Penicillinase (EC 3.5.2.6)	pH	Physical Soluble	1–2 weeks 3 weeks	0.5–2 min 2 min	400 1000	10^{-2}–10^{-4} 10^{-2}–10^{-4}
10. Uric acid	Uricase (EC 1.7.3.3)	Pt(O_2)	Chemical	4 months	30 sec	10	10^{-2}–10^{-4}
11. Amygdalin	β-Glucosidase (EC 3.2.1.21)	CN^-	Physical	3 days[g]	10–20 min	100	10^{-2}–10^{-5}
12. Cholesterol terol	Cholesterol oxidase (EC 1.1.3.7)	Pt(H_2O_2)	Soluble		2 min		10^{-2}–10^{-4}

(*Continued*)

Table 23. (*Continued*)

Type	Enzyme	Sensor	Immobilization[a]	Stability	Response time	Amount of enzyme (U)	Range (mol/1)[b]
13. Phosphatase	Phosphatase/ glucose oxidase (EC 3.1.3.1/1.1.3.4)	$Pt(O_2)$	Chemical	4 months	1 min	10	10^{-2}–10^{-4}
14. Nitrate	Nitrate reductase/nitrate reductase (EC 1.9.6.1/ 1.6.6.4)	NH_4^+	Soluble		2–3 min	10	10^{-2}–10^{-4}
15. Nitrate	Nitrite reductase (EC 1.6.6.4)	NH_3(gas)	Chemical	3–4 months	2–3 min	10	5×10^{-2}–10^{-4}
16. Sulphate	Aryl sulphatase (EC 3.1.6.1)	Pt	Chemical	1 month	1 min	10	10^{-1}–10^{-4}

[a] "Physical" refers to polyacrylamide gel entrapment in all cases; "chemical" is attachment chemically with glutaraldehyde to albumin, to polyacrylic acid, or to acrylamide, followed by physical entrapment.
[b] Analytically useful range, either linear or with reasonable change if curvature is observed.
[c] "Kinetic", rate of change in current measured after 12 sec; "steady-state" current reaches a maximum in 1 min.
[d] Preparation lacks stability as evidenced by constant decrease in signal each day.
[e] Electrode responds to L-cysteine, L-leucine, L-tyrosine, L-tryptophan, L-phenylalanine, and L-methionine.
[f] Electrode responds to D-phenylalanine, D-alanine, D-valine, D-methionine, D-leucine, D-norleucine, and D-isoleucine.
[g] Time required for signal to return to base line before re-use.

Table 24. The main characteristics of enzyme electrodes in biomedical studies. (From Makarov K.A., and Kibardin S.A., Immobilized biopreparations in medicine (Russ.), Moscow, Medicine, 1980, with permission)

Enzyme	Material under determination	Immobilization method	What is determined by sensor	Electrode
Asparaginase	L-asparagine	Entrapment into gel	NH_4^+	Platinum
Asparaginase	L-asparagine	Entrapment into membrane	NH_4^+ crystalline	"—"
β-Glucosidase	Amigdaline	Entrapment into gel	CN^-	dialysis
β-Glucosidase	Amigdaline	Entrapment into gel and coating with film	CN^-	"—"
Glutaminase	L-Glutamine	Entrapment into space between electrode and membrane	NH_4^+ pH-electrode	Cation
Glucose oxidase	Glucose	Entrapment into gel	O_2	Oxygen
" — "	"—"	Entrapment into space between electrode and membrane	Hydroquinone	Platinum
Glucose oxidase	Glucose	Covalent binding	H_2O_2	Platinum
"—"	"—"	Entrapment into gel	H_2O_2	"—"
"—"	"—"	Entrapment into space between electrode and membrane	H_2O_2	"—"
"—"	"—"	Entrapment into polyacrylamide or ethyleneglycolmethacrylate gel	H_2O_2	"—"
Lactatdehydrogenase	Lactic acid	Entrapment between electrode and membrane	—	Platinum
L-aminoacid oxidase	L-phenylalanine	Entrapment into gel	NH_4^+	Glass
"—"	"—"	Covalent binding		
Urease	Urea	Entrapment into gel	NH_4^+	Glass
"—"	"—"	Entrapment into gel and film coating	NH_4^+	"—"
"—"	"—"	Entrapment into gel	NH_4^+	"—"
"—"	"—"	"—"	NH_4^+	"—"

High sensitivity and operational stability are the main advantages of enzyme electrodes. Thus, according to [707] electrodes with covalently bound enzymes can be used to perform 200–1000 analyses during a relatively long period of time.

One of the most developed methods is the determination of glucose in biological fluids. For this purpose, several types of enzyme electrodes have been developed; all of them use glucose oxidase as biochemical component. Glucose is determined with an electrode containing the enzyme immobilized in polyacrylamide gel, and the electrochemical sensor is polarographic oxygen electrode which determines the glucose concentration based on oxygen consumption measurements. Glucose has also been determined with platinum electrodes measuring the amount of hydrogen peroxide formed by the immobilized glucose

Table 25. Possible electrode sensors useful in construction of enzyme electrode. (From Guilbalt, G.G. and de Olivera Neto, G., in Immobilized Cells and Enzymes. A Practical Approach, Woodward, J., Ed., IRL Press, Oxford, 1985, 55, wtih permission)

	Useful for
A. Potentiometric sensors	
NH_3	Urea, amino acids, glutamine, glutamic acid, nitrate, nitrite, creatine, lyase and deaminase enzymes
CO_2	Urea, amino acids, decarboxylative enzyme systems
pH	Penicillin, RNA, DNA, glucose, enzyme reactions giving pH change
I^-	Glucose, amino acids, cholesterol, alcohols
CN^-	Amygdalin
B. Voltammetric (amperometric) sensors	
O_2	Glucose, amino acids, organic acids, alcohols, uric acid, cholesterol, phosphate, all O_2 consuming enzymes
Pt or C	All redox enzymes, sulphate, uric acid, cholesterol, alcohols glucose, amino acids, NADH/NADPH systems.

oxydase. A hydroquinone-sensitive platinum electrode is a variety of the latter method. Platinum electrodes are also used where the enzyme had been incorporated into the space between the electrode and membrane; into ethyleneglycol metaclytate gel; immobilized on rotating disc electrode [709–721].

Urea has been determined with electrodes containing immobilized urease and ammonia-selective electrochemical elements, since urea is hydrolysed to ammonia by urease. As in the case of glucose-sensitive electrodes, there are several electrodes available, which differ by the method of enzyme immobilization and by the location of the enzyme and the electrochemical component. The difference in optimal pH values for the electrode (ammonia evolves from the solution at alkaline pH) and for urease (7.0–7.4) is a considerable neatness of traditional urea-sensitive enzyme electrodes [722–724].

Immobilized uricase incorporated in platinum electrodes has been used to determine uric acid (which is oxidized by this enzyme to hydrogen peroxidase and allantoin in the presence of oxygen) [726]. Cholesterol can be determined in the presence of immobilized cholesterol oxidase, H_2O_2 being one of the products of enzymatic reaction that is recorded polarographically or potentiometrically by ion selective electrodes [722, 727].

The possibility of simple and rapid determination of penicillins in biological fluids with immobilized penicillinase (which catalyses penicilline hydrolysis to phenylacetic and 6-aminopenicillanic acids) exists. The acid generated can be measured by changes in the concentration of hydrogen ions by means of a standard glass electrode. Usually, penicillinase immobilized on polymeric gel is applied on the electrode surface [728].

Recently, more metabolites are being determined by enzyme electrodes, among them ascorbic acid, determined with immobilized ascorbate oxidase

[729–731]. An amperometric oxygen sensor with a dialysis membrane containing catalase immobilized via 2,4-dichloro-6-metoxy-s-triasine is used for hydrogen peroxide determination [732]. The response time of this electrode is less than 12 sec and the linear response to hydrogen peroxide is in the concentration range from 10^{-3} to 10^{-5} mol/l. The determination of hydrogen peroxide is also of particular importance for the analysis of other metabolites, (in which H_2O_2 is a product of enzymatic conversion) and the development of rapid and sensitive H_2O_2-specific enzyme electrodes is a very important task. Immobilized oxalate oxidase, in combination with potentiometric CO_2-sensors or amperometric H_2O_2-sensors, has been used for the determination of oxalate in urine [733]. A bioamperometric method of arachidonic acid determination (the metabolic precursor of prostaglandins) in biological fluids has also been developed [734], using lactate oxidase immobilized on acetyl cellulose membranes. Oxygen electrode with modified enzyme determine blood phenylalanine for the diagnosis of phenylketonuria [734]. Enzyme electrodes, containing immobilized creatine aminohydrolase, creatine amido hydrolase and sarcosine oxidase, have been developed for the determination of creatine and creatinine in the blood [735, 736]. Measurements can be performed in 20 μl volumes of serum and the electrodes remain stable for over 9 months at 4 °C.

Electrodes with immobilized cofactors (e.g., flavine) have been proposed for the determination of physiologically important, cofactor-dependent enzymes [737–739]. Studies to develop new modifications of enzyme electrodes are in progress. Thus, an electrode with urease immobilized on polyester cloth and an ammonia selective membrane has been described [740]. An interesting approach has been developed in [741] where the enzyme electrode is obtained by incorporation of an enzyme into an electroconductive polymer, sedimented on the surface of the platinum electrode. The authors of [742] have developed a new type of enzyme electrode, where ferrocene derivatives serve as mediators between an enzyme and the electrode. The authors of [743] described a combined method for glucose determination which possesses some features of a continuous flow system with immobilized enzyme, but provides the electrochemical determination of the product. In this case, the enzyme (glucose oxidase) was covalently coupled to an antibody and introduced into the device containing immobilized antigen. This resulted in enzyme immobilization by an antigen-antibody bridge. When a glucose containing sample is introduced into the device, glucose forms hydrogen peroxide, which in turn is registered by an appropriate sensor.

It stands to reason that these few examples can not cover all the studies on enzyme electrodes, but nevertheless they demonstrate how rapid progress in this field is.

6.2 *Enzyme Immunoassay*

The property of antigens to bind specific antibodies is the basis of numerous immunochemical assays which are widely used in various fields. Many of them involve the use of antigens or antibodies labeled with specific markers which help to determine the presence of an analyzed compound with maximal sensitivity.

Enzymes are of particular importance among such markers. They proved to be suitable to detect minute amounts of metabolites and convenient for histochemical studies and the quantitative determination of antigens and antibodies. The many different immunochemical assays are now united under the separate heading of "enzyme immunoassay" or EIA.

Numerous reviews and books [744–748, 748a] indicate that currently, EIA is used for the determination of a wide spectrum of substances: antibodies, peptide and steroid hormones, pharmacological preparations, viral and bacterial antigens, pesticides, and many others. Heterogenous (solid-phase) and homogenous types of EIA have been developed, depending on the methods used for separation of components in the immunochemical reaction [744]. Solid-phase methods are based on antibodies and antigens immobilized on insoluble carriers, while homogenous methods are based on the effect of the antibody-dependent modulation of the enzyme activity (or cofactor) which is used as antigen label.

Apparently, the main problem of EIA is the incorporation of the enzyme label into antigen and antibody molecules, i.e. enzyme immobilization on antigens or antibodies. Currently, there are only few generally accepted marker enzyes for EIA (see Table 26). Such a limited set is explained by strict requirements for marker enzymes: they must be highly active under the conditions of the assay; the enzyme reaction (its products or substrates) must be determined with maximal simplicity; enzymes should fully retain their activity when conjugated with antigens or antibodies; the samples investigated should contain no free enzyme, its inhibitors or activators; and finally, it is desirable that the particular enzyme should be available in a highly purified form and should not be very costly.

A number of methods and reagents for the binding of marker enzymes to antigen or antibodies exists: from the simplest bifunctional reagent – glutaric

Table 26. Enzyme used in enzyme immunoassay

Enzyme	Mol. weight	Substances
Heterogenous EIA		
Peroxidase	40,000	H_2O_2; o-dianizidin, 5-amino-salicylic acid, o-phenylene-diamine
Alkaline phosphatase	100,000	n-nitrophenylphosphate
β-D-Galactosidase	518,000	O-nitrophenyl, β-D-galactoside
Glucose oxidase	150,000	glucose, O_2
Glucoamylase	90,000	dextrin
Acetylcholine esterase	250,000	acetylcholine
Homogenous EIA		
Lysozyme	14,000	cell walls
Malatedehydrogenase	70,000	malate, NAD
β-D-Galactosidase	518,000	O-nitrophenyl, β-D-galactoside
Glucose-6-phosphate dehydrogenase	102,000	Glucose-6-phosphate, NAD

dialdehyde, which has been used in the first steps of EIA development, to more complex reagents used today. As a rule, protein binding to ligands occurs via protein amino and sulfhydril groups. In the case of peroxidase, preliminary oxidation of carbohydrate fragments of the enzyme molecules with the introduction of free aldehyde groups (which bind to ligand amino groups via Schiff bases) is widely used. The principal methods of enzyme conjugation with antibodies and antigens via bifunctional reagents are presented in Table 27.

Immunoenzyme conjugates with peroxidase [750], alkaline phosphatase [751], glucose oxidase [752] and glucose amilase [753] have been obtained by glutaric aldehyde coupling. The composition of the conjugates can be varied by changing the aldehyde concentration. An increase in the concentration of cross-linking agents brought about the increase in the molecular weight of the immunoglobulin-enzyme conjugate.

4,4-Difluor-3,3-dinitrophenylsulfon interacts mainly with lysine NH_2-groups, but tyrosine and histidine residues can also enter the reaction when coupling is performed at pH 10–11. The reagent can be used only for a one-step

Table 27. Reagents for immunoenzyme conjugates synthesis (E-enzyme; P-antigen or antibody) [749]

N	Reagent	Conjugate structure
1	Glutaraldehyde 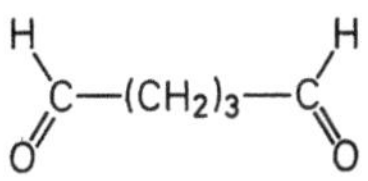	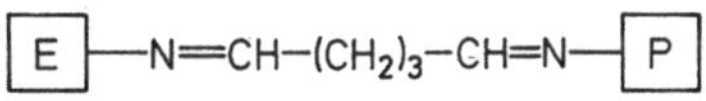
2	4, 4′-difluoro-3, 3′-dinitrodiphenylsulfone 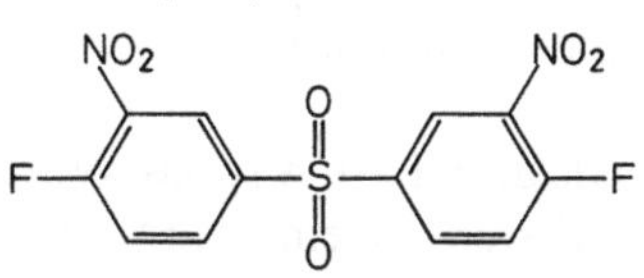	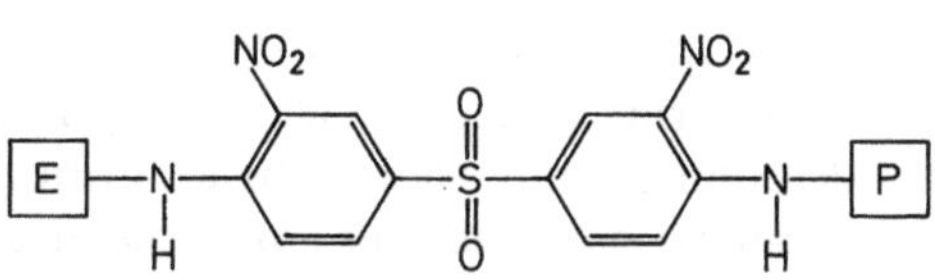
3	Cyanuric chloride 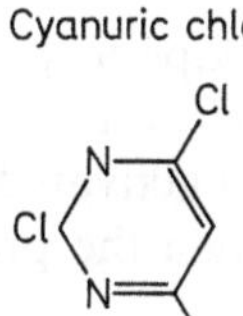	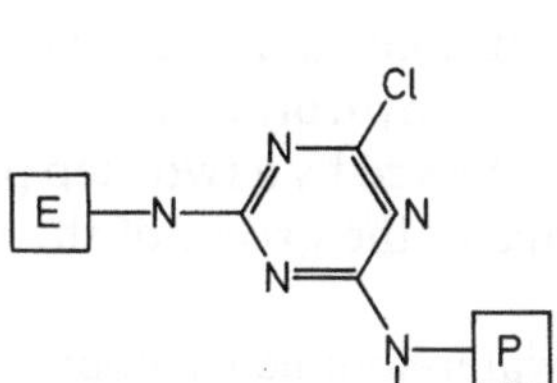
4	Water soluble carbodiimides R—N=C=N—R′	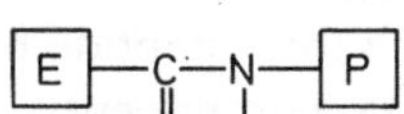

(Continued)

Table 27. (*Continued*)

N	Reagent	Conjugate structure
5	2, 4′-diisocyanatotoluene 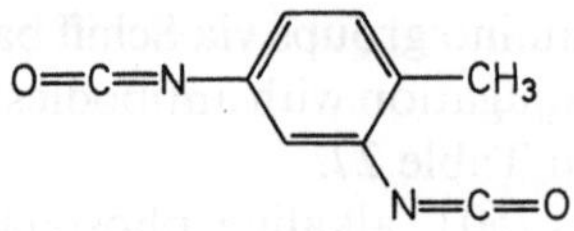	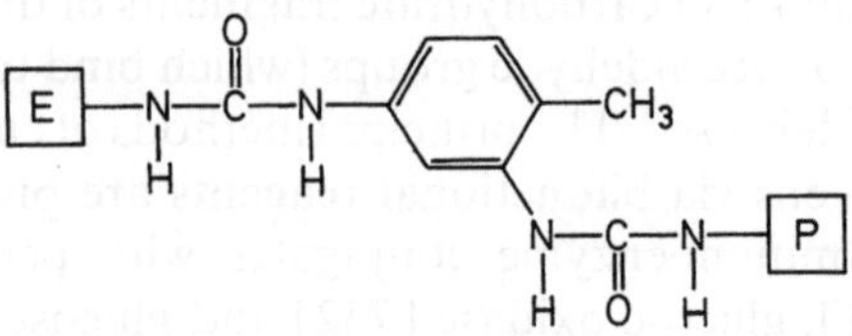
6	N, N-o-phenylmaleimide 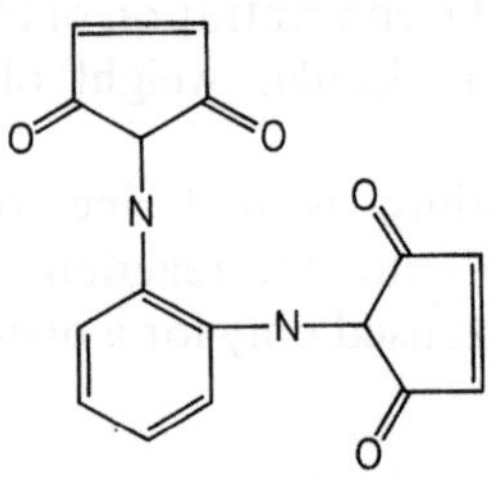	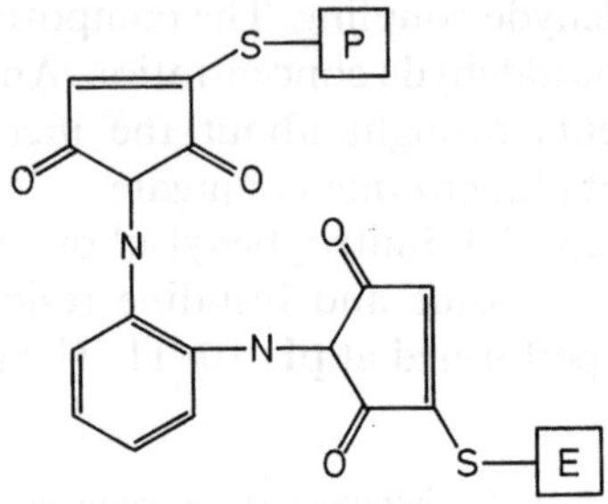
7	Sodiam periodate ($NaJO_4$) (enzyme-peroxidase)	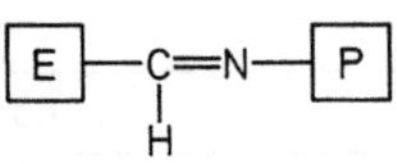
8	p-benzoquinone 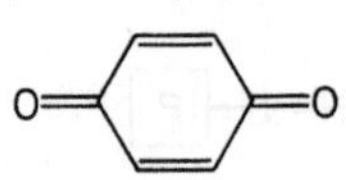	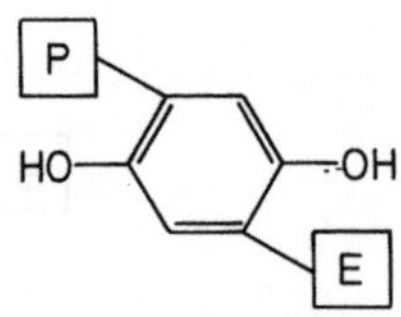

synthesis (proteins and cross-linking agents are mixed simultaneously), since the active fluorine molecules are rapidly hydrolyzed.

Protein binding via cyanuric chloride proceeds at pH 10. A rapid hydrolysis of cyanuric chloride rules out the use of multi-step schemes for conjugate synthesis.

2,4-Diisocyanatotoluene reacts mainly with lysine NH_2-groups in protein molecules. The rate of hydrolysis in aqeuous medium is slow, therefore, conjugate synthesis can be achieved by a two-step process: the enzyme is first activated by a cross-linking reagent, the excess of the latter is removed and then the protein added.

Among the water-soluble carbodiimides, the most suitable compounds for the synthesis of immunoenzyme conjugates are 1-ethyl-3-(3-dimethylamino-propyl)meto-toluene sulfonate and 1-cyclohexyl-3-2-(morpholinoethyl)meto-toluene sulfonate, which form peptide bonds between free carboxy groups and NH_2-groups of protein components. The stability and activity of hybrid molecules obtained by this method are usually found to be decreased when compared with the initial components. A possible reason of this phenomenon is the carbodiimide interaction with various functional groups of protein molecules that leads to conformational changes accompanied by a decrease in specific activity [754].

The methods developed for immuno enzyme conjugates with β-galactosidase are also based on the use of SH-groups of the enzyme, the blocking of which has no effect on the catalytical properties [755]. IgG reduced by mercaptoethylamine or its Fab fragments are bound to β-galactosidase via N,N'-o-phenylenedimaleimide, which specifically reacts with protein SH-groups. Conjugate yield measured by enzyme activity is 50%; the components almost completely retain their specific properties. Sulfhydryl groups can be introduced under mild conditions into compounds containing amino groups. Insulin-β-galactosidase conjugates were obtained by introduction of SH-groups into insulin [756]. Introduction of maleimide groups into insulin is another method of obtaining the same conjugates [757].

p-Benzoqinone reacts mainly with NH_2- and SH-groups of proteins and with carbohydrate residues. The reaction is performed in two steps, at neutral or low alkaline pH. A study on antibody-peroxidase conjugates based on this method indicates that their composition is highly homogenous and that 50–80% of activity is retained in the hybrid molecules [758].

The synthesis of conjugates of a defined composition is a very important task. In this connection, the method for producing homogenous antibody conjugates with invertase is of particular importance [753]. This method is based on affinity immobilization of invertase on a concanavalin A-Sepharose column with subsequent enzyme modification by glutaryc dialdehyde. The conjugate is eluted after application of antibodies on the column with bound enzymes and the formation of covalent bonds between protein and enzyme; the invertase bond with Sepharose is splitted by α-methylmannoside. The conjugate obtained contains about 90% of hybrid molecules with a components ratio of 1:1.

Several other methods of enzyme immobilization on antibodies have been introduced, such as the use of metamaleimidobenzoyl-N-oxysuccinimide ester, a hetero-bifunctional reagent [760]. This reagent was used for conjugating insulin with β-D-galactosidase. Glucose oxidase was also conjugated with antibodies via the N-oxysuccinimide ester N-(4-carboxycyclohexylmethyl)maleimide [761]. Enzyme activity decreased by 20%. N-succinimidyl-3-(2-pyridildithio)-propionate became the most popular among hetero-bifunctional reagents. It permitted the binding of peroxidase with immunoglobulins into homogenous complexes, the yield being 60% [762]. Recently, conjugates of glucose-6-phosphate dehydrogenase with low molecular weight haptens, cortisol and progesteron, have been described [763]. The enzyme binds 20–25 molecules of the hormone and retains 40–50% of its initial activity. The conjugates obtained are suitable of the EIA of these hormones. Rased on maleimide coupling, a method for the determination of elcatonin by EIA, using β-galactosidase, has been developed [764].

Immunological methods of labelling antibodies with enzymes are also of particular interest [765]. These methods are based on the formation of an antigen-antibody bond with the enzyme serving a label.

One of the approaches is called the method of "hybrid antibodies". Fab-fragments of antibodies against a specific antigen and an enzyme are obtained by enzymatic hydrolysis. Then the mixture of the hydrolysis products is reduced by mercaptoethanol and the Fab-fragments are reversibly dissociated into symmetrical components. After removal of the reducing agent, the molecules dissociate and form hybrid molecules of antibodies, specific towards the specified antigen, and enzyme. Antibody-enzyme complex is formed after addition of an appropriate enzyme.

Enzyme conjugates with Staphylococcus protein A form a stable complex with Fc-fragments of antibodies from various animal species [741, 766]. A method which combines covalent and immunochemical conjugates is based on this phenomenon. It permits the binding of antisera with a low antibody titer.

The shelf-life of immunoenzyme conjugates is a very important property for practical use in medicine. Universal methods of conjugates stabilization have not been yet developed. There is also no correlation between the stability of conjugates and the method of their preparation. Nevertheless, the stability of hybrid molecular exceeds considerably the stability of antigens and antibodies labeled with radioactive isotopes. Enzyme conjugates do not change their properties when stored in lyophilized state for two years. Preparations containing alkaline phosphatase, β-galactosidase, lysozyme or peroxidase appear to be stable in solution (at +4 °C) for more than one year [749, 767, 768].

The development of methods for immunoadsorbent synthesis is of particular importance, mainly concerning solid-phase EIA with immobilized antigens and antibodies for the rapid and quantitative separation of analytes. Immobilization is usually achieved either by protein covalent coupling with a polymeric matrix or by adsorption on a carrier. Cellulose, Sepharose, Sephadex, Biogel, Agarose, porous glass and silochrom are usually used as carriers. Insoluble derivatives of antibodies can be obtained with cross-linking reagents, such as glutaryc dialdehyde, water-soluble carbodiimides, isothiocyanate, etc. [744, 749].

There are several types of EIA; one of them being a competitive method where a specific antigen and a known amount of enzyme-labeled antigen are added to immobilized antibodies. The concentration of the enzyme label, measured in the solution after achieving the equilibrium, is proportional to the concentration of the substance determined. Another method with double antibodies is based on a competitive assay in homogenous phase with subsequent addition of an immunoadsorbent containing an excess of antiidiotypic antibodies. A third approach is a "sandwich" method, in which the assayed solution containing the antigen studied is added to the immunoadsorbed antibody. After equilibrium has been established, unbound components are washed out and enzyme-labeled antibodies are added into the system. After the reaction reaches equilibrium for the second time, the excess of antibody-enzyme conjugate is removed and the quantity of carrier-bound enzymatic label (which is proportional to the quantity of the antigen to be determined) is measured. Another methods of analysis also exist.

The use of cofactors (ATP and NAD) as antigen or antibody markers is also of great interest. Complexes of this type are usually used in reactions involving luciferase and dehydrogenase [769]. The methods are based on the "shielding" of antigen-bound cofactors by antigens, which results in the inhibition of the enzymatic reaction. When the analyzed antigen is added to the system in a free state, it competes with the cofactor-labeled antigen for binding sites on the antibody molecule and causes partial dissociation of the antibody: (antigen/cofactor) complex and an increase in the enzymatic activity of the solution. The latter can be registered by conventional methods and used for the quantitative estimation of the analyzed antigen in the sample.

The use of immunoenzyme electrodes combines their simplicity and sensitivity with the specificity of immunoadsorption [770]. Antibodies are usually immobilized on nylon nets surrounding Clark's oxygen electrode. Catalase (or glucose oxidase) labeled antigen and free antigen compete for the antibody binding sites. After removal of unbound components and addition of enzymatic reaction substrates, the electrode current is proportional to the concentration of oxygen released or consumed in the catalytic reaction with the antigen-bound enzyme. The duration of analysis is 10 to 15 minutes, the sensitivity (for albumin and insulin) 10^{-5} g/ml. The potential of immunoenzyme electrodes is very high because the activity of many different enzymes can be measured electrochemically. Some parameters of enzymes, which can be used in the method described, are presented in Table 28.

A new version of enzyme immunoassay has been developed, based on measurements of heat effect on enzymatic reactions, the enzyme being part of an immunoenzyme conjugate with a substrate [771]. Antibodies against human serum albumin have been covalently bound with BrCN-Sepharose. Then, a mixture of human serum albumin and human serum albumin-catalase conjugate is analyzed on the column with the adsorbent. After washing the free component

Table 28. List of useful enzymes for electrochemical enzyme immunoassays. (From Sittampalam G.S. and Wilson G.S., Trends in Analytical Chemistry, 3, No 4, 96, 1984, with permission)

Enzyme	Specific activity[a]	Electro-active species detected	Method of detection
Glucose oxidase (1.1.3.4)	1.7×10^5 (pH 5.1, 35 °C)	H_2O_2	Amperometry
L-Amino acid oxidase	6.0×10^3 (pH 6.5, 37 °C)	H_2O_2, NH_3	Amperometry (H_2O_2) Potentiometry (NH_3)
Amine oxidase (1.4.3.6)	1.5×10^2 pH 7.2, 37 °C	H_2O_2, NH_3	Amperometry (H_2O_2) Potentiometry (NH_3)
Catalase (1.11.1.6)	4.0×10^7 (pH 7.0, 25 °C)	O_2	Amperometry
Glucose-6-phosphate dehydrogenase (1.1.1.49)	4.0×10^5 (pH 7.4, 25 °C)	NADH	Amperometry
Urease (3.5.1.5)	1.1×10^6 (pH 7.0, 25 °C)	NH_3	Potentiometry

[a] Micromoles of substrate converted per minute per gram of protein, at optimum pH and temperature shown in parentheses. Adapted from specific activity information given in "Sigma" Catalog, 1983.

from the column, a H_2O_2 solution is applied and the heat effect of the reaction (using an appropriate thermistor) which is proportional to the initial albumin concentration is registered. The analysis can be performed within 10 minutes, the sensitivity being 10^{-10} M. The method can also be used for the determination of low molecular weight compounds, such as hydrogen peroxide (catalase is used as enzymatic component of the registration system), phenols (tyrosinase is the enzymatic component), glucose (using glucose oxidase) and sucrose (using invertase) [772].

In all the above EIA methods, the enzyme is used as an antigen or antibody label and its distribution between solid and liquid phases depends on the concentration of the substance to be measured. Currently, new EIA systems are being developed, based on enzyme activation as a result of antigen-antibody specific interaction. One of the possible ways to develop such systems is enzyme entrapment into liposomes with an appropriate antigen bound on their surface. The addition of antibodies and a complement-containing solution into the system results in the formation of antigen-antibody complexes on the liposomes with the consequent lysis of the latter by action of the complement. Lysis is accompanied with the release of intraliposomal enzymes into the solution. The addition of a sample containing the specific antigen into the system inhibits this lysis and decreases the amount of antibody and complement molecules bound to the liposomes. The rate increase of enzyme activity in the solution correlates with the initial concentration of the antigen. The time required for theophylline determination in such a system (liposomes with peroxidase) is 15 min, sensitivity

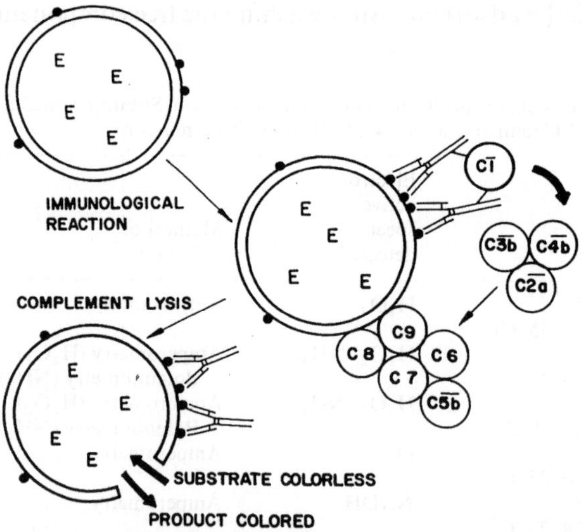

Fig. 59. Principle of total complement activity test. The *concentric circles* represent the lipid bilayer membrane of the liposome; (●), DNP antigens attached to the surface; *E*, alkaline phosphatase encapsulated in the liposome; *Cn*, different components of the complement system (Reproduced with permission from Bowden DW et al. (1986) Clin Chem 32:275)

Table 29. Sensitivity of enzyme immunoassay towards different antigens

Antigen	Enzyme	Sensitivity limit $\times 10^{12}$M
Immunoglobulin G	β-galactosidase	2
Immunoglobulin A	alkaline phosphatase	50
Insulin	β-galactosidase	20
Choriogonadotropin	peroxidase	80
Tyroxin	malatedehydrogenase	26×10^4
Progesteron	β-galactosidase	160
Phenobarbital	glucose-6-phosphate dehydrogenase	4×10^6
Digoxin	glucose-6-phosphate dehydrogenase	640
Surface antigen of hepatitis B virus	peroxidase	0, 9
Cancer-embryonic antigen	alkaline phosphatase	25

is 4×10^{-9} M with electrochemical registration of the peroxidase activity in the reaction of NADH oxidation [773]. Not only enzymes, but also fluorescent and spin-active compounds can be entrapped into liposomes (for the scheme of the method, see Fig. 59).

Highly sensitive chemoluminiscent and fluorometric methods of peroxidase and β-galactosidase detection have also been developed. The optimization of the enzymatic reaction conditions plays an important role. For example, the rate of peroxidase-catalyzed oxidation of o-dianisidine and o-phenylendiamine increases in the presence of non-ionic detergents which permits to increase EIA sensitivity [774]. Table 29 contains some data on the sensitivity of determination for various compounds by solid phase methods of EIA.

EIA have greatly contributed to the development of methods for the detection of antibodies to various antigens in the blood. EIA modifications have been already developed for the diagnostics of alveococcosis, brucellosis, syphilis, helmintosis, toxoplasmosis, sleeping sickness, measles, malaria etc. [749].

Automated systems have been designed to perform analyses of more than 1000 serum samples per day.

6.3 Immobilized Bioluminescent Systems

Bioluminescence is a result of luciferin oxidation under the action of luciferase. These names include a whole class of substrates and enzymes which differ in structure and ways of forming excited products in different organisms. Bioluminescent reactions have the following distinctive features: high quantum yield (from 0.1 to 1.0); strict specificity; absolute dependence on ATP. The high quantum yield causes the extreme sensitivity of luciferase systems: 1 quant of light is formed per each reacted molecule of substrate.

As has been pointed out in [775, 776], bioluminescent analytical systems are used for determination of many important metabolites including substrates of ATP- or NADH-dependent enzymes. Many other compounds can be determined using systems of conjugated reactions. Since the use of native enzyme (usually enzymes purified from insects) have a lot of drawbacks, an attempt has been made to obtain immobilized bioluminescent systems and to use them for analysis. In [777] it has been shown that immobilized luciferase exhibits high stability on storage and under the conditions of the analysis, lower sensitivity towards inhibitors and possible repeated use.

Numerous studies have demonstrated that immobilized luciferases obtained with polysaccharide carriers (cellulose, dextran or agarose) activated by cyanogen bromide, periodate oxidation or cyanuric chloride, exhibited optimal properties [777–779]. The properties of various preparations of immobilized luciferase are presented in Table 30 (from [776]). Naturally, luciferase in the sensors used in practice has not been immobilized on Sepharose or Sephadex which

Table 30. Preparations of immobilized luciferase [776]

Enzyme source	Carrier	Activating agent	Sensitivity limit	Stability
Raw extract	Sepharose	Cyanogen-bromide	50 nM	at 25 °C, in 6,5 nM dithiotreitol $\tau_{1/2} = 100$ hrs
Crystalline luciferase	Alkylaminated glass	Glutaraldehyde	—	—
Raw extract	Ultradex	Cyanogen-bromide	50 nM	at 25 °C, in 6,5 mM dithiotreitol $\tau_{1/2} = 10$ min
	Ultragel		0,1uM	$\tau_{1/2} = 5$ min
	Cellophane		0,5 uM	$\tau_{1/2} = 20$ hrs
Raw extract	Sepharose	Cyanogen-bromide	0,1 nM	at 4 °C lyophilized preparation, containing dithiotreitol, Mg^{2+} salts and luciferin, does not lose activity during 6 months
Raw extract	Cellophane film	Sodium periodate	0,1 nM	at 25 °C, in suspension with 1 mg/ml dithiotreitol and 0,1 M Mg^{2+} $\tau_{1/2} = 10$ min
		Cyanuric chloride	0,1 nM	$\tau_{1/2} = 5$ hrs
Crystalline luciferase	CL-Sepharose 4B	Cyanogen-bromide	0,15 nM	at 20 °C, in 0,1M phosphate buffer with 15% glycerol and 0,1 nM dithiotreitol; freezing thawing without enzyme activity loss

[a] $\tau_{1/2}$-enzyme half-inactivation time.

are difficult to separate from the reaction medium. Immobilized luciferases most commonly used are incorporated into membranes, rods and column reactors. Thus, luciferase has been immobilized on dialysis cellulose tubings that can be reused repeatedly [780, 781]. Bienzyme systems of light-emitting bacteria were incorporated into albumin gel, cross-linked with glutaryc aldehyde and obtained as discs, each being suitable for 100 measurements [782]. The use of immobilized

Table 31. ATP determination as an analytical method in microbiology, medicine and science [776]

Destination	Sphere of application
Quantitative measurement of microbial cells, algae and fungi (sensitivity limit 10^4 cells)	Diagnostics of microbial infections in urine and blood
Measurement of microbial cultures growth rate	Determination of bacteria sensitivity towards antibiotics
	Determination of antibiotics concentration in blood serum following inhibition of microbial cells growth
Measurement of ATP accumulation rate in the medium	Investigation of cellular membranes damage following quantity of released ATP (determination of red blood cells and platelet cytolysis during blood storage; determination of different cells viability)
Measurement of absolute ATP concentration and its changes	Immunocofactor analysis utilizing ATP as a label
	Determination of ATP-dependent enzymes activity
	Determination of substrates of ATP-dependent enzymes in clinical analysis (creatine phosphate, cAMP, pyrophosphate)

Table 32. Analytical applications of immobilized bioluminescent enzymes (From Kricka L.J. and Thorpe G.H.G., Analyst, 108, 1274, 1983, with permisstion)

Immobilized enzymes	Analyte
Firefly luciferase	ATP
Bacterial luciferase	$FMNH_2$
NAD(P)H:FMN oxidoreductase, bacterial luciferase	NAD(P)H
Alcohol dehydrogenase, NADH: FMN oxidoreductase, bacterial luciferase	Ethanol
Glucose-6-phosphate dehydrogenase, NADH: FMN oxidoreductase, bacterial luciferase	Glucose-6-phosphate
Alanine dehydrogenase, NADH: FMN oxidoreductase, bacterial luciferase	L-Alanine
Glutamate dehydrogenase, NADH: FMN oxidoreductase, bacterial luciferase	L-Glutamate
Malate dehydrogenase, NADH: FMN oxidoreducause, bacterial luciferase	L-Malate
Lactate dehydrogenase, NADH: FMN oxidoreductase, bacterial luciferase	L-Lactate, NAD
6-Phosphogluconate dehydrogenase, NADPH: FMN oxidoreductase, bacterial luciferase	6-Phosphogluconate, NADP
Hexokinase, glucose-6-phosphate dehydrogenase, NADPH: FMN oxidoreductase, bacterial luciferase	D-Glucose

luciferase in a flow-through system in automated analysis has also been described [783].

The determination of the ATP content in various biological systems is one of the most important fields of application of bioluminescent analysis. ATP in the medium points to the presence of living cells i.e. its determination can serve as analytical method for the discovery of various microbial contaminations. Besides, the ATP content in cells characterizes their viability to some extent. Possible fields of ATP clinical application are presented in Table 31.

Immobilized bacterial luciferase was used for the analytical determination of glucose-6-phosphate [784]. The same analysis was performed in a flowing column using bacterial luciferase immobilized on BrCN-activated Sepharose [785]. Creatine phosphokinase, the marker of myocardial infarction, was determined by bacterial luciferase immobilized in albumin gel cross-linked with glutaryc dialdehyde [782]. Other possibilities of analytical use of immobilized bioluminescent systems are presented in Table 32.

6.4 Devices with Immobilized Enzymes; Miscellaneous

Various structures of analytical devices with immobilized enzymes have been proposed: tubular, column and bobbins. In tubular reactors, the enzyme is covalently immobilized on the inner surface of nylon tubes, its length varying from 1 to 3 m. Enzymes immobilized on granules of an insoluble carrier or entrapped in the particles of polymeric gel are used in column reactors. Enzymes incorporated into fibers (e.g., triacetate cellulose fibers) are used in bobbin reactors. The flow rate of the mixture of reagents and samples with the analyte is such that when the mixture passes through enzymatic reactor, the reaction is either terminated (final product analysis) or proceeds to a certain degree of conversion (fixed-time analysis).

Currently, many enzymatic methods of analysis have been automated. Two main types of automatized analyzers are being used: the Technicon flow analyzer, in which a certain flow rate of reagent supply is chosen, thus determining the duration of the reaction [707]; and the Jemsec centrifuge analyzer, in which the mixture of the reagents and the substance to be determined is automatically dispersed into separate sections of spectrophotometric cuvettes, the solution is rapidly mixed during centrifugation and the optical density in all cuvettes is registered simultaneously.

Methods based on the use of the above analytical devices serve for the determination of many metabolites: glucose, urea, uric acid, cholesterol, triglycerides, amino acids, ascorbate, lactate, pyruvate and drugs or toxins.

To have a few examples: a thermistor placed into the device with urease immobilized on aminosilan glass was successfully used for the determination of urea [786]. Urea can also be determined by ammonia formation, its accumulation being registered spectrophotometrically while the analysed solution passes through the continuous flow reactor, containing urease incorporated into cellulose triacetate fibers or immobilized on the inner surface of nylon tubes [787, 788]. An interesting sensor of a mixed type is described in [789], where the authors used double systems for the determination of urea and creatinine. The system consist of immobilized urease or creatinine deaminase and nitrating bacteria which in the first case convert ammonia to nitrate and in the second also utilize ammonia formed from creatinine together with N-methyl hydantoin.

Spectrophotometric and luminescent methods of determination of uric acid are based on the use of conjugated reactions. The amount of hydrogen peroxide formed from uric acid upon its oxidation in the presence of immobilized uricase has been measured [790, 731]. The use of uricase immobilized in the reactor in the form of nylon tubes has also been reported. In this case, the enzyme was bound to a partially hydrolyzed nylon surface via glutaryc dialdehyde [792]. In a continuous flow glass reactor, uricase was immobilized on activated silica fibers also via glutaryc dialdehyde [793]. Uricase stabilized by protamine was used for uric acid determination, with uricase immobilized on protamine-coated glass beads [794]. An automated system of uric acid determinationin in the blood has been developed, based on the immobilized enzyme stable even at 70 °C. Usually, the H_2O_2 formed as a result of the reaction is registered colorimetrically. The process of hydrogen peroxide determination by aldehyde dehydrogenase immobilized in nylon tubes has also been described [795].

Similar methods have been applied for glucose determination in biological solutions. Flow systems are used where the amount of glucose is determined by the amount of hydrogen peroxide formed from glucose by glucose oxidase. Hydrogen peroxide is registered by chemoluminescence using luminol [796]. Hydrogen peroxide can be also determined enzymatically using peroxidase and chromogenic substrates [797]. The use of a flow reactor with glucose oxidase immobilized via glutaryc dialdehyde on alkylamino-glass beads also provied to be successful [799].

Many examples of the use of reactors with immobilized enzymes for the determination of clinically significant metabolites and exogenous compounds can be demonstrated. They are based on well-known enzymatic reactions and the problem in each particular case is the choice of the optimal construction of the reactor and optimal parameters for its functioning. The analysis of some other more exotic metabolites is worth mentioning. Thus, coimmobilization of 3-α-oxysteroid dehydrogenase and dihydrolipamide reductase on cellulose cyclic imidocarbonate provides a continuous flow system for the determination of such important compounds as 3-α-oxysteroids [800]. β-Fructofuranosidase, immobilized on concanavalin A and sorbit dehydrogenase were used for the spectrophotometric insulin determination [801]. A multi-channel flow analyzer has been proposed in [802]. It can be used for many immobilized enzymes or bienzyme systems. Such metabolites as uric acid, cholesterol and its esters, glutamate, lactate, ammonia, glucose, etc. can be determined by chemoluminescence.

Although analytical methods with immobilized enzymes are continuously being developed, alternate trends exist to develop must simpler approaches.

Table 33. Some methods of determination of organic and inorganic compounds in clinical analysis using immobilized enzymes; adapted from [722]

Analyte	Enzymes used	Determination limits, working intervals, experimental errors	Carrier for enzyme immobilization	Stability of enzymatic sensor or reagent	Notes
Indicator reaction: urea $\rightarrow NH_3 + CO_2$					
Urea	Urease	0.1–50 nM	polyacrylamide gel or glass	500 analyses during 60 days	suspension in nylon cover; response time 2–3 min
—	—	0.05–10 mM	albumin film crosslinked with glutar, aldehyde	20 analyses per hour, 200–1000 analyses with one sensor	enzyme membrane on electrode surface
Indicator reactions: a) urea $\rightarrow NH_3 + CO_2$; b) NH_3 determination following Bertelote reaction					
Urea	Urease	0.1–10 mM	nylon tube	4 months, 200 analyses with one tube	autoanalyzer Technicon
Indicator reaction: Uric acid $+ O_2 \rightarrow$ allantoin $+ CO_2 + H_2O_2$					
Uric acid	Uricase	0.1–100 mM	glass	3 months	O_2-electrode response time 60 sec
Indicator reactions: a) uric acid $+ O_2 \rightarrow$ allantoin $+ CO_2 + H_2O_2$; b) H_2O_2 + aminophenasone + dichlorophenol $\rightarrow$ dye					
Uric acid	a) uricase b) peroxidase	3–75 uM	nylon tube	4000 analyses during 3 months with one tube	autoanalyser Technicon
Indicator reactions: L-aminoacid $+ O_2 \rightarrow$ ketoacid $+ NH_4 + H_2O_2$					
L-Amino acids	L-aminooxidase	0.01–10 mM	membrane	4 months	O_2-electrode
L-Phenylalanine	L-aminooxidase	0.1–1 mM	membrane	2 weeks	NH_3-electrode; response time 1–2 min

Indicator reactions: a) $RCH(NH_2)COOH \rightarrow CO_2 + RCH_2NH_2$: b) $RCH_2NH_2 + O_2 + H_2O \rightarrow RCHO + NH_3 + H_2O_2$					
L-arginine	a) L-arginine decarboxylase b) diamine oxidase	0.02–0.2 mM error 3–5%	albumin film	—	O_2-electrode; response time 1–2 min
Indicator reactions: a) L-aspartate + 2-oxoglutarate → oxaloacetate + L-glutamate b) oxaloacetate + NAD × H_2 ⇄ L-malate + NAD					
L-aspartic acid	a) aspartate aminotrans-ferase b) malate dehydrogenase	0.1–2 mM	Sepharose	more than 24 h under working conditions at 37 °C	columne 0.7 ml
Indicator reaction: lactate + NAD ⇄ pyruvate + NAD × H_2					
Lactate	lactate dehydrogenase	0.1–10 mM	Sepharose	40 analyses during 14 days	LKB-analyzer
Indicator reaction: glucose + O_2 → gluconic acid + H_2O_2					
Glucose	glucose oxidase	0.03–1 mM	Silochrom	2 months 200 analyses during 2 weeks	O_2-electrode columne with immobilized enzyme
—	—	1–30 mM	nylon tube	10 thousands analyses with 1 immobilized enzyme stock	autoanalyzer Technicon, O_2-electrode
Indicator reactions: a) glucose + O_2 → gluconic acid + H_2O_2; b) $H_2O_2 \rightarrow H_2O + O_2$					
Glucose	a) glucose oxidase	0.01–0.9 mM error 1–2%	glass	30 days	columne with thermister, 12–30 analyses per hour
Indicator reactions: a) glucose + O_2 → gluconic acid + H_2O_2; b) homovanillic acid + H_2O_2 → chromophore					
Glucose	a) glucose oxidase b) peroxidase	2–20 mM, error 3%	BrCN-cellulose	several months, hundreds of analysis	sample 5 um

(*Continued*)

Table 33. (*Continued*)

Analyte	Enzymes used	Determination limits, working intervals, experimental errors	Carrier for enzyme immobilization	Stability of enzymatic sensor or reagent	Notes
Indicator reactions: a) glucose $+ O_2 \rightarrow$ glyconic acid $+ H_2O_2$; b) H_2O_2 + hydrazon-3-methyl-2-benzothiazolinon + dimethylaniline → dye					
Glucose	a) glucose oxidase b) peroxidase	3–30 mM, error 1–2%	nylon tube	25 thousands analyses during 74 days with the same tube	autoanalyzer Technicon
Indicator reactions: a) glucose + ATP → glucose-6-phosphate + ADP; b) glucose-6-phosphate + NADP → 6-phosphoglucolactone + NADP × H_2					
Glucose	a) hexokinase b) glucose-6-phosphate dehydrogenase	1–10 mM,	NH_2-substituted nylon tube	3 500 analyses per 90 hours	autoanalyzer Technicon
Indicator reaction: glucose + NAD → glucosolacton + NAD × H_2					
Glucose	glucose dehydrogenase	2.5–13 mM	nylon tube	30 analyses per hour	autoanalyzer Technicon
Indicator reactions: a) sucrose → α-D-glucose + D-fructose; b) α-D-glucose → β-D-glucose; c) β-D-glucose $+ O_2 \rightarrow$ D-glucose-δ-lactone					
Sucrose	a) invertase b)mutarotase c) glucose oxidase	0.05–10 mM, error $< 7\%$	collagen film crosslinked with glytaraldehyde	85 analyzes during 10 days	O_2-electrode coated with membrane, response time 2–3 min
Indicator reaction: D-galactose $+ O_2 \rightarrow$ galactohexodialdose $+ H_2O_2$					
Galactose	galactose oxidase	5–3000 mg, error 3%	polycarbonate membrane	more than a month	H_2O_2-electrode with trilayer membrane, response time 40 sec

Indicator reactions: a) cholesterol $+ O_2 \rightarrow$ cholesterol-4-on-3 $+ H_2O_2$; b) $H_2O_2 \rightarrow O_2 + H_2O$					
Cholestetol (free)	a) cholesterol oxidase b) catalase	0.01–0.1 mM	porous glass	more stable than in solution	column with termister
Indicator reactions: a) cholesterol ester $\rightarrow$ cholesterol; b) cholesterol $+ O_2 \rightarrow$ cholesterol-4-on-3 $+ H_2O_2$; c) H_2O_2 + aryloxalate $\rightarrow$ product + hv					
Cholesterol (total)	a) cholesterol hydrolase b) cholesterol oxidase	1 nM–0.3 mM error 5%	foam quartz spheres	—	column, analysis time 15–20 min
Indicator reaction: penicillin $\rightarrow$ phenylacetic acid + 6-aminopenicillanic acid					
Penicilline and analogues	penicillinase	0.01–3 mM, error 3–5%	1) enzyme is entrapped in the space between electrode surface and coating film 2) polyacrylamide gel 3) glass disc	300 analyses during 6 weeks with the same electrode	electrodes (1) and (2) are sensitive to Me^+, electrode (3) is not
Indicator reaction: a) penicillin $\rightarrow$ phenylacetic acid + 6-aminopenicillanic acid; b) 6-aminopenicillanic acid $+ I_2$ (in starch solution) $\rightarrow$ decolorizing					
Penicillin	penicillinase	0.02–1 mM	nylon tube	50 days of non-stop working	flow autoanalyzer, 12000 analyses during 14 days with the same tube
Indicator reactions: a) apopyruvate ocidase $+ Zn^2 \rightarrow$ active enzyme; b) pyruvate $+ O_2$ + phosphate $\rightarrow$ acetyl phosphate $+ H_2O_2$; c) H_2O_2 + bis-(3,4,6-trichlorophenyl)-oxalate $\rightarrow$ hv + product					
Zn^{2+}	pyruvate kinase	8 pg	foam quartz spheres	20–25 days	enzymatic flow reactor

Thus, in [803] a nylon brush with cellulose triacetate fibers containing immobilized lactate dehydrogenase was described. The brush is intended for the clinical analysis of pyruvate. For this purpose, the sample is agitated with the brush for 10 min with subsequent analyte registration. The brush can be re-used for many analyses within 3 months, its activity remaining unchanged. The same device based on fibers with immobilized urease has also been described for ammonia determination. In this case, the ammonia formed was determined by the Bertelot method after the sample has been agitated by an enzymatic brush for a specified time [804].

Thus, even the incomplete data so far presented (some of them are shown in Table 33) demonstrate that the field of clinical laboratory analysis is one of the most promising for immobilized enzymes application. Taking into consideration that the use of immobilized enzymes in clinical analysis does not require prolonged biomedical trials (in contrast to immobilized enzymes as drugs), one may assume that this field in medicine will continue to develop very rapidly.

7 Conclusion

I hope that the reviewed material in this book will help the reader to understand the main problems and possibilities of application of immobilized enzymes in medicine. We did not elaborate on the use of immobilized enzymes for the production of therapeutic drugs. In general, the main approaches to the application of immobilized enzymes are well defined. It has become accepted that preparations and materials with immobilized enzymes are required for practical medicine and need to be developed. The future of this trend is very promising and the first results of clinical trials of various preparations have justified the hopes. If this book provided and facilitated orientation in the numerous data on medical aspects of enzyme immobilization, the author will consider his goal to be fulfilled.

8 References

1. Wolf M, Ransberger K (1972) Enzyme therapy, Vantage Press, New York
2. Holcenberg JS, Roberts J (eds) (1981) Enzyme as drugs, John Wiley, New York
3. Holcenberg JS (1982) Ann Rev Biochem 51: 795
4. Tager JM, Hooghwinkel GJM, Daems WT (eds) (1974) Proceedings of the workshop on cell biological and enzymological aspects of the therapy of lysosomal storage diseases, Leiden, The Netherlands, Elsevier, New York
5. Tager JM, Daems WT, Hoodhwinkel GJM (eds) (1974) Enzyme therapy in lysosomal storage diseases, North/Holland, Amsterdam
6. Grabowsky GA, Desnick RJ (1981) Enzyme replacement in genetic diseases. In: Holcenberg JS, Roberts J (eds) Enzymes as drugs, John Wiley, New York, p 167
7. Graham DY (1981) Enzyme therapy of digestive disorders. In: Holcenberg JS, Roberts J (eds) Enzymes as Drugs. John Wiley and Sons, New York, p 331
8. Capizzi RL, Cheng Y-C (1981) Therapy of neoplasia with asparaginase. In: Holcenberg JS, Roberts J (eds) Enzymes as drugs. John Wiley and Sons, New York, p 1
9. Holcenberg JS (1981) Therapy of neoplasia with other nonessential amino acid degrading enzymes. In: Holcenberg JS, Roberts J (eds) Enzymes as drugs. John Wiley and Sons, New York, p 25
10. Roberts J (1981) Therapy of neoplasia by deprivation of essential amino acids. In: Holcenberg JS, Roberts J (eds) Enzyme as drugs. John Wiley and Sons, New York, p 63
11. Kalghatgi KK, Bertino JR (1981) Folate-degrading enzymes: a review with special emphasis on carboxypeptidase G. In: Holcenberg JS, Roberts J (eds) Enzyme as drugs. John Wiley and Sons, New York, p 77
12. Levy CC, Karpetsky TP (1981) Human ribonucleases. In: Holcenberg JS, Roberts J (eds) Enzyme as drugs. John Wiley and Sons, New York, p 103
13. Schmidt B, Zahn R (1980) Exonuclease und diese enthaltendes cytostatisches Mittel, FRG Patent 2145528
14. Lazerson J (1981) Clotting factor replacement. In: Holcenberg JS, Roberts J (eds) Enzyme as drugs. John Wiley and Sons, New York, p 241
15. Verstraete M (1979) Pathol Biol 23: 239
16. Marsh N (1981) Fibrinolysis, John Wiley, Chichester
17. Collen D, Lijnen HR (1986) CRC Crit Rev Oncol Hematol 4: 249
18. Fletcher AP, Alkjaersing NK (1981) Fibrinolytic and defibrinating enzymes. In: Holcenberg JS, Roberts J (eds) Enzymes as drugs. John Wiley and Sons, New York, p 209
19. Bloom AI, Thomas DP (1981) Haemostasis and thrombosis, Churchill Livingstone, Edinburgh
20. Lijnen HR (1984) Fibrinolysis: molecular mechanism and pathophysiological aspects. Sangre 29: 755
21. Lijnen HR, Collen D (1986) Molecular mechanisms of thrombolytic therapy. 16 suppl 3: 3
22. Matsuo O, Bando H, Okada K, Tanaka K, Tsukada M, Iga Y, Arimura H (1986) Thrombolytic effect of single-chain pro-urokinase in a rabbit jugular vein thrombosis model. Thromb Res 42: 187
23. Borrelli F, Antonett F, Martelli F, Carpino L (1986) The co-operative action of hyaluronidase and urokinase on the isoproterenol-induced myocardial infarction in rats. Thromb Res 42: 153

24. Williams DO, Borer J, Braunwald E, Chesebro JH, Cohen LS, Dalen J, Dodge HT, Fransis CK, Knatterud G, Ludbrook P, Markis JE, Mueller H, Desvingne-Nichens P, Passamani ER, Powers ER, Rao AK, Roberts R, Ross A, Ryan TJ, Sobel B, Winniford M, Zaret B (1986) Intravenous recombinant tissue-type plasminogen activator in patients with acute myocardial infarction: a report from the NHLBI thrombolysis in myocardial infarction trial. Circulation 73: 338
25. Myers ML, Belli R, Lekich RF, Hartley CJ, Roberts R (1985) Enhancement of recovery of myocardial function by oxygen free-radical scavengers after reversible regional ischemia. Circulation 72: 915
26. Sawada T (1984) Method for treatment of arthrosis deformans with elastase. US Patent 4446192
27. Fujisaki S, Mitani M (1984) Pronase used for the treatment of diseases of the liver and kidneys in humans and animals. US Patent 4485095
28. Proceedings of IV USSR conference on medical enzymology. Astrakhan 1984
29. Pinnell SR (1985) Method for the prevention and treatment of scars with enzymes. US Patent 4524065
30. Ohinshi H, Kosuzume H, Suzuki Y, Mochida E (1982) Agent therapeutique a base d'une protease acid, pour le traitement des troubles allergiques, des maladies avec immunocomplexe et des tumeurs. French Patent 2499405
31. Wilder MS (1986) Procede de prevention et de soulagement d'irritations cutanees avec une formule contenant de la dismutase de superoxyde. International Patent WO 86/025 56
32. Zaborsky OR (1973) Immobilized enzymes. CRC Press, Cleveland
33. Mosbach K (ed) (1976) Immobilized enzymes. Meth Enzymol 44
34. Triven MD (1980) Immobilized enzymes. An introduction and applications in biotechnology. John Wiley and Sons New York
35. Torchilin VP, Martinek K (1979) Enzymes stabilization without carriers. Enzyme microb Technol 1: 174
36. Berezin IV, Martinek K (eds) (1982) Introductory course of applied enzymology. Immobilized enzymes. MGU Moscow
37. Matiassen B (ed) (1983) Immobilized enzymes, cells and organells. v. 1–2, CRC Press, Boka Raton
38. Mosbach K (ed) (1988) Immobilized Enzymes. Meth Enzymol v. 1–2, p 117
39. Ugarova NN (1982) Stabilization of enzymes. In: Berezin IV, Martinek K (eds) Introductory course of applied enzymology. Immobilized enzymes MGU Moscow p. 156
40. Mozaev VV (1982) Inactivation and reactivation of proteins (enzymes). In: Berezin IV, Martinek K (eds) Introductory course of applied enzymology. Immobilized enzymes MGU Moscow, p 180
41. Martinek K, Torchilin VP, Siksnis VA, Maksimenko AV, Smirnov VN, Berezin IV (1980) The mechanism of stabilizing action of salts in the process of proteolytic enzymes thermoinactivation. DAN SSSR (Russ) 251 (5): 1169
42. Trubetskoy VS, Torchilin VP (1986) Artificial and natural thermostabilization of subunit enzymes. Do they have similar mechanism? Int J Biochem 17: 661
43. Torchilin VP, Maksimenko AV, Smirnov VN, Berezin IV, Klibanov AM, Martinek K (1979) The principles of enzyme stabilization. IV. Modification of "key" functional groups in the tertiary structure of proteins. Biochim Biophys Acta 567: 1
44. Torchilin VP, Maksimenko AV, Smirnov VN, Berezin IV, Klibanov AM, Martinek K (1978) The principles of enzyme stabilization. III. The effect of the length of intra-molecular cross-linkages on thermostability of enzymes. Biochim Biophys Acta 522: 277
45. Korshak VV, Shtilman MI (1984) Polymers in immobilization and modification of Natural compounds. Nauka, Moscow
46. Dunkan R, Kopecek J (1984) Soluble synthetic polymers as potential drug carriers. Adv Polym Sci 57: 51
47. Drobnik J, Rypacek F (1984) Soluble synthetic polymers in biological systems. Adv Polym Sci 57: 1
48. Lloyd JB (1985) Macromolecules as vehicles for intracellular drug delivery. Pharm Int 10: 252
49. Kopecek J, Rejmanova P (1983) Enzymatically degradable bonds in synthetic polymers. In: Bruck SD (ed) Controlled drug delivery. CRC Press, Boka Raton v. 1: 81
50. Poznansky MJ, Juliano RL (1984) Biological approaches to the controlled delivery of drugs: a critical review. Pharm Rev 36: 277

51. Lloyd JB, Dunkan R, Kopecek J (1980) Synthetic polymers as carriers for chemotherapeutic agents. Biochem Soc Trans 14: 391
52. Plate NA, Valuev LI, Chupov VV (1984) The chemistry of macromonomers based on physiologically active substances and some problems of the synthesis of polymers for biomedical application. Pure Appl Chem 56: 1351
53. Schacht E, Ruys L, Vermeersch J, Remon JP (1984) Polymer-drug combinations: synthesis and characterization of modified polysaccharides containing procainamide moieties. J Contr Rel 1: 33
54. Holmberg K, Hyden H (1985) Methods of immobilization of proteins to polymethylmethacrylate. Prep Bioch 15: 309
55. Kopecek J (1984) Controlled biodegradability of polymers—a key to drug delivery systems. Biomaterials 5: 19
56. Dunkan R, Lloyd JB, Kopecek J (1980) Degradation of side chain of N-(2-hydroxypropyl) methacrylamide copolymers by lysosomal enzymes. Biochem Biophys Res Commun 94: 284
57. Dunkan R, Rejmamova P, Kopecek J, Lloyd JB (1981) Pinocytic uptake and intracellular degradation of N-(2-hydroxypropyl) methacrylamide copolymers. A potential drug delivery system. Biochim Biophys Acta 678: 143
58. Lloyd JB, Dunkan R, Kopecek J, Rejmanova P (1984) Targeting and lysosomal handling of polymethacrylamide-oilgopeptide conjugates. In: Gregoriadis G, Poste G, Senior J, Trouet A (eds) Recetor-mediated trageting of drugs. Plenum Press, New York
59. Rejmanova P, Kopecek J, Pohl J, Baudys M, Kostka V (1983) Polymers containing enzymatically degradable bonds. Degradation of oilgopeptide sequences in N-(2-hydroxypropyl) methacrylamide copolymers by bovine spleen cathepsin B. Makromol Chem 184: 2009
60. Cartlidge SA, Dunkan R, Lloyd JB, Kopeckova-Rejmanova P, Kopecek J (1987) Soluble, crosslinked N-(2-hydroxypropyl) methacrylamide copolymers as potential drug carriers. 2. Effect of molecular weight on blood clearance and body distribution in the rat after intravenous administration. Distribution of unfractionated copolymer after intraperitoneal, subcutaneous or oral administration. J Contr Rel 4: 253
61. Cartlidge SA, Dunkan R, Lloyd JB, Kopeckova-Rejmanova P, Kopecek J (1987) 3. Trageting by incorporation of galactosamine residues. Effect of route of administration. J Contr Rel 4: 265
62. Leong KW, Kost J, Mathiowitz E, Langer R (1986) Polyanhydrides for controlled release of bioactive agents. Biomaterials 7: 364
63. Controlled drug delivery. Bruck SD (ed) (1983) vol. 1–2. CRC Press, Boka Raton
64. Hoffman AS (1982) Synthetic polymer biomaterials in medicine. In: Benoit H, Rempp P (eds) Macromolecules. Pergamon Press, Oxford, p 321
65. Torchilin VP, Mazaev AV, Il'ina EV, Goldmacher VS, Smirnov VN, Chazov EI (1980) Chemical aspects of enzyme stabilization and modification for use in therapy. In: Wingard LB, Berezin IV, Klyosov AA (eds) Future directions for enzyme engineering. Plenum Press, New York, p 219
66. Larionova NI, Torchilin VP (1980) Current state and prospects of using in medicine of immobilized physiologically active substances of protein nature (Review) Khim Farm J (Russ) 14: 21
67. Larionova NI, Torchilin VP (1982) Application of immobilized physiologically active substances of protein nature in medicine. In: Berezin IV, Martinek K (eds) Introduction into applied enzymology. MGU Moscow, p 284
68. Korshak VV, Torchilin VP, Shtil'man MI, Il'ina EV, Brudz SP (1983) On the relative reactivity of some water soluble epoxy-containing polymers towards α-chymotrypsin. DAN SSSR (Russ) 273: 627
69. Maksimenko AV, Nadirashili LA, Abramova VV, Erkomaishvili GS, Katsarava RG, Torchilin VP (1989) The modification of soluble complex of papaya proteinases with synthetic polymers. J Contr Rel 10: 131
70. Marshall JJ (1975) Preparation and characterization of a dextran-amylase conjugate. Carbohydr Res 49: 389
71. Marshall JJ, Rabinowitz ML (1975) Enzyme stabilization by covalent attachment of carbohydrate. Arch Biochem Biophys 167: 777
72. Marshall JJ, Rabinowitz ML (1976) Stabilization of catalase by covalent attachment to dextran. Biotechnol Bioeng 18: 1325

73. Marshall JJ, Humphreys JD (1979) Synthesis of a soluble catalase-dextran conjugate. Biotechnol Bioeng 19: 1739
74. Marshall JJ, Rabinowitz ML (1976) Preparation and characterization of a dextran-trypsin conjugate. J Biol Chem 251: 1081
75. Marshall JJ, Humphreys JD, Abramson SL (1977) Attachment of carbohydrate to enzymes increases their circulatory lifetimes. FEBS Lett 83: 249
76. Torchilin VP, Reizer IL, Smirnov VN, Chazov EI (1976) Enzyme immobilization on biocompatible carriers. III. Immobilization of α-chymotrypsin on soluble dextran. Bioorg Khim (Russ) 2: 1252
77. Torchilin VP, Tischenko EG, Smirnov VN, Chazov EI (1977) Immobilization of enzymes on slowly soluble carriers. J Biomed Mater Res 11: 223
78. Marshall JJ (1978) Manipulation of the properties of enzymes by covalent attachment of carbohydrate. TIBS 3: 79
79. Marshall JJ, Humphreys JD (1979) Experimental enzyme therapy: suppression of allergic reactions in preimmunized animals by administration of exogenous enzymes in the form of enzyme—dextran conjugates. J Appl Biochem 1: 88
80. Sherwood RF, Baird JK, Atkinson T, Wiblin CN, Rutter DA, Ellwood DC (1977) Enhanced plasma persistence of therapeutic enzymes by coupling to soluble dextran. Biochem J 164: 461
81. Jakunitskaya LM, Samartsev MA, Tarasova GV, Martynov VF (1980) Preparation of subtilisine BPN' and dextrane conjugates. Prikladn Biokhim Mikrobiol (Russ.) 16: 232
82. Blomhoff HK, Christensen TB (1983) Effect of dextran and dextran modifications on the thermal and proteolytic stability of conjugated bovine testis β-galactosidase and human serum albumin. Biochim Biophys Acta 743: 401
83. Maksimenko AV, Konovalova OYu, Berdichevsky VP, Arkhipova OG, Torchilin VP In vitro and in vivo studies of native and dextran-modified hyaluronidase. Bull Exper Biol Med (Russ) 102: 567
84. Arkhipova OG, Yaglov VV, Maksimenko AV, Pavlovskaya LV, Konovalova OYu, Varsanovich EA, Bezrukavnikova LM, Fedorova VI, Fedorova VN, Torchilin VP (1987) The effect of dextran-modified hyaluronidase in experimental silicosis. Bull Exper Biol Med (Russ) 103: 221
85. Abuchowski A, McCoy JR, Palczuk NC, van Es T, Davis FF (1977) Effect of covalent attachment of polyethylene glycol on immunogenicity and circulating life of bovine liver catalase. J Biol Chem 252: 3582
86. Abuchowski A, van Es T, Palczuk NC, Chen R, Davis FF (1977) Effect of covalently attached polyethylene glycol on the immunogenicity of enzymes. Fed Proc Fed Am Soc Exp Biol 36: 867
87. Abuchowski A, Kazo G, Verhoest J (1984) Antitumor properties of polyethylene glycol-asparaginase conjugates. Cancer Biochem Biophys 7: 175
88. Davis FF, Abuchowski A, van Es T, Palczuk NC (1979) Soluble, non-antigenic polyethylene glycol-bound enzymes. Am Chem Soc Polym Prep 20: 357
89. McDonald RJ, Berger EM, White CW, White JG, Freeman BA, Repine JE (1985) Effect of superoxide dismutase encapsulated in liposomes or conjugated with polyethylene glycol on neutrophil bactericidal activity in vitro and bacterial clearance in vivo. Amer Rev Resp Dis 131: 633
90. Boccu E, Velo G, Veronese F (1982) Pharmacokinetic properties of polyethylene glycol derivatized superoxide dismutase. Pharm Res Communs 14 2 113
91. Koide A, Kobayashi S (1983) Modification of amino groups in porcine pancratic elastase with polyethylene glycol in relation to binding ability towards anti-serum and to enzymic activity. Biochem Biophys Res Commun 111: 659
92. Savoca KV, Abuchowski A, van Es T, Davis FF, Palczuk NC (1979) Preparation of a non-immunogenic arginase by the covalent attachment of polyethylent glycol. Biochim Biophys Acta 578: 47
93. Abuchowski A, Davis FF (1981) Soluble polymer-enzyme adducts. In: Holcenberg JS, Roberts J (eds) Enzymes as drugs. John Wiley and Sons, New York, p 367
93a. Hershfield MS, Buckley RH, Greenberg ML, Melton AL, Schiff R, Hatem C, Kurtzberg J, Markert ML, Kobayashi RH, Kobayashi AL, Abuchowski A (1987) Treatment of adenosine deaminase deficiency with polyethylene glycol-modified adenosine deaminase. N Engl J Med 316: 589
94. Ashihara Y, Kone T, Yamazaki S (1978) Modification of E coli L-asparaginase with polyethylene glycol: disappearance of binding ability to asparaginase serum. Biochem Biophys Res Commouns 83 2 385

95. Lisi P, Es T, Abuchowski A (1982) Enzyme therapy I Polyethylene glycol β-glucoronidase conjugates as potential therapeutic agents in acid mucopolysaccharidosis. J Appl Biochim 4: 19
96. Wieder KJ, Davis FF (1983) Enzyme therapy: II Effect of covalent attachment of polyethylene glycol on biomedical parameters and immunological determinants of β-glucosidase and α-galactosidase. J Appl Biochem 5: 337
97. Lindenbaum GM, Mirgorodskaya OA, Moskvichev BV, Tereshin IM (1987) Study of some physicochemical and enzymatic properties of dextran based polymeric derivatives of terrilytin. Khim Farm Zhurn (Russ) 11: 81
98. Bogacheva TI, Mirgorodskaya OA, Grinmberg GE, Mikhailets GA, Moskvichev BV, Tereshin IM (1981) Effect of chemical modification on proteinase interaction with inhibitors, coagulability and acute toxicity. Biokhimia (Russ) 46: 863
99. Kashkin AP, Lindenbaum GM, Mahovenko LV, Konshina IZ, Moskvichev BV, Tereshin IM (1978) Antigenic properties of proteolytic enzyme, terrilytin modified with dextran. Khim Pharm J (Russ) 12(7) 27
100. Lindenbaum GM, Yakovleva MD, Mirgorodskaya OA, Moskvichev BV, Tereshin IM (1982) Chemical modification of terrilytin with water-soluble oxidized dextrans. Prikl Biokhim Mikrobiol (Russ) 18(2): 212
101. Specht BU, Wahl M, Kobb, HJ, Brendel WW (1975) Application of polyvinylpyrrolidone as a carrier for kallikrein. Arch Int Pharmacodyn Ther 213: 242
102. Geiger B, von Specht BU, Arnon R (1977) Stabilization of human beta-D-acetylhexosaminidase a towards proteolytic inactivation by coupling it to poly(N-vinylpirrolidone) Eur J Biochem 73: 141
103. Von Specht BU, Brendel W (1977) Preparation and properties of trypsin and chymotrypsin coupled covalently to poly(N-vinylpirrolidone) Biochim Biophys Acta 484: 109
104. Tennikova TB, Moskvichev BV, Samsonov GV (1980) Study of physico-chemical properties of terrilytin modified with a copolymer based on vinylpyrrolidone. Biokhimia (Russ) 45: 438
105. Tennikova TB, Panarin EF, Mirgorodskaya OA, Samsonov GV, Moskvichev BV (1977) Immobilization of proteolytic enzyme terrilytin on water soluble polymeric matrix. Khim Pharm J 7: 86
106. Maksimenko AV, Walsmann P (1981) Herstellung und Charakterisierung eines loslichen Ocrase—Polyacrylamid—praparates. Pharmazie 36: 117
107. Brown RK, Trzpis MA, Sela M, Anfinsen CB (1963) Studies on the antigenic structure of ribonuclease. J Biol Chem 238: 3876
108. Arnon R, Neurath H (1970) Immunochemical studies on bovine trypsin and trypsinogen derivatives. Immunochemistry 7: 241
109. Uren JR, Ragin RC (1927, 1979) Improvement in the therapeutic, immunological and clearance properties of Escherichia coli and Erwinia carotovora L-asparaginases by attachment of poly-DL-alanyl peptides. Cancer Res, p 39
110. Kurinenko BM, Belyaeva MI, Cherepneva IE, Kupriyanova–Ashina FG (1977) Permeability of dextran-bound ribonuclease through the vessel barrier and tumor cell membrane. Voprosy Onkologii (Oncol Probl Russ) 23(5): 86
111. Kurinenko BM, Belyaeva MI, Cherepneva IE, Viesture ZA (1977) The antitumor effect of Ser. Marcescens rullease bound covalently to soluble dextran. Voprosy Onkologii (Onkol Probl Russ) 23: 93
112. Keturkene AP, Astrauskas VI (1983) Effect of the administration of immobilized enzymes on collagen degradation in rats with adjuvant arthritis. In: Debov SS (ed) Proceedings of the VI all-union symposium on medical enzymology. Inst Exper Clin Med (Russ) Vilnus p 121
113. Abuchowski A, van Es T, Palczuk NC, Davis FF (1977) Alteration of immunological properties of bovine serum albumin by covalent attachment of polyethylene glycol. J Biol Chem 252: 3578
113a. Wie SI, Wie CW, Lee WY, Filion LG, Sehon AH, Akerblom E (1981) Suppression of reaginic antibodies with modified allergens. III. Preparation of tolerogenic conjugates of common allergens with monomethoxypolyethylene glycols of different molecular weights by the mixed anhydride method. Int Archs allergy appl Immun 64: 84
113b. Lee WY, Sehon AH, Akerblom E (1981) Suppression of reagenic antibodies with modified allergens. IV. Induction of suppressor T cells by conjugates of polyethylene glycol (PEG) and monomethoxy PEG with ovalbumin. Int Archs allergy appl Immun 64: 100
114. Liu F-T, Zinnecker M, Hamaoka T, Katz DH (1979) New procedures for preparation and isolation of conjugates of proteins and a synthetic copolymer of D-amino acids and immunochemical characterization of such conjugates. Biochemistry 18: 690

115. Kabanov VA, Petrov RV, Khaitov RM (1982) A new principle for preparation of artificial immunogens. Zhurn V KH O (Russ) 27: 57
116. Makela O, Peterty F, Outschoorn IG, Richter AW, Seppala I (1984) Immunogenic properties of α(1 → 6)dextran, its protein conjugates and conjugates of its breakdown products in mice. Scand J Immunol 19: 541
117. Rozenberg GYa (1978) Chemically modified hemoglobin is an artificial oxygen carrier. DAN SSSR (Russ) 243: 1320
118. Tam S-C, Wong JT-F (1980) Modification of hemoglobin upon covalent coupling to dextran-enhanced stability against acid denaturation and reduced affinity for haptoglobin. Can J Biochem 58: 732
119. Tam S-C, Blumenstein J, Wong JT-F (1978) Blood replacement in dogs by dextran-hemoglobin. Can J Biochem 56: 681
120. Dellacherie E, Bonneaux F, Labrude P, Vigneron C (1983) Modification of human hemoglobin by covalent association with soluble dextran. Biochim Biophys Acta 749: 106
121. Guillochon D, Esclade L, Remy MH, Thomas D (1981) Studies on hemoglobin immobilized by cross-linking with glutaraldehyde. Cross-linked soluble polymers and artificial membranes. Biochim Biophys Acta 670: 332
122. Larionova NI, Kazanskaya NR, Sacharov IYu, Mityushina GV (1980) Soluble high-molecular weight derivatives of pancreatic trypsin inhibitor; Modification of pancreatic trypsin inhibitor by soluble polysaccharides activated by titanium tetrachloride. Biokhimiya (Biochemistry, Russ) 45: 638
123. Larionova NI, Mityushina GV, Kazanskaya NF, Blidchenko YuA, Berezin IV (1984) Carbohydrate-containing derivatives of the trypsin-kallikrein inhibitor aprotinin from bovine organs. I. Modification with lactose, characterization and behavior of the preparation in vivo. Hoppe-Seyler's Z Physiol Chem 365: 791
124. Larionova NI, Mityushina GV, Kazanskaya NF, Blidchenko VA, Berezin IV (1985) Carbohydrate-containing derivatives of the trypsin-kallikrein inhibitor aprotinin from bovine organs. II. Inhibitor coupled to the (carboxymethyl)dextran derivatives of D-galactose. Biol Chem Hoppe-Seyler 366: 743
125. Odya CE, Levin Y, Erdos EG, Robinson CJG (1978) Soluble dextran complexes of kallikrein, bradykinin and enzyme inhibitors. Biochem Pharm 27: 173
126. Venter JC (1982) Immobilized and insolubilized drugs, hormones, and neurotransmitters: properties, machinsms of action and applications. Pharmacol Rev 34: 153
127. Barashkova GM, Prussakov AN, Polosatov MV, Klimov PK (1978) The effect of synthetic macromolecular gastrin upon functioning gastric glands. Sechenov Physiol J USSR 64: 1348
128. Torchilin VP, Martinek K (1979) Enzyme stabilization without carriers. Microbial Technol 1: 74
129. Torchilin VP, Maksimenko AV, Smirnov VN, Berezin IV, Klibanov AM, Martinek K (1978) The principles of enzyme stabilization. III. The effect of the length of intra-molecular cross-linkages on thermostability of enzymes. Biochim Biophys Acta 522: 277
130. Torchilin VP, Trubetskoy VS, Omel'yanenko VG, Martinek K (1989) Stabilization of subunit enzyme by intersubunit crosslinking with bifunctional reagents. Studies with glyceraldehyde-3-phosphate dehydrogenase. J Mol Catal 19: 291
131. Torchilin VP (1983) Immobilized enzymes and the use of immobilization principles for drug targeting. In: Goldberg EP (ed) Targeted drugs. John Wiley and Sons, New York, p 127
132. Venkatasubramanian K, Vieth WR, Bernath FR (1974) Use of collagen immobilized enzymes in blood treatment. Enzyme engineering 2: 439
133. Holcenberg JS, Schmer G, Teller DC, Roberts J (1975) Biologic and physical properties of succinylated and glycosylated Acinetobacter glutaminase-asparaginase. J Biol Chem 250: 4165
134. Dillon JC, Wade CWR, Daly WH (1976) Enzyme immobilization on fibrin biotechnol Bioeng 18: 133
135. Snyder PD, Wold F, Bernlohr RW, Dullum D, Desnick RJ, Krivit W, Condie RM (1974) Enzyme therapy. II. Purified human β-galactosidase A. Stabilization to heat and protease degradation by complexing with antibody and by chemical modification. Biochim Biophys Acta 350: 432
136. Poznansky MJ, Shaudling M, Salkie MA, Elliot J, Lan E (1982) Advantages in the use of L-asparaginase-albumin polymer as an antitumor agent: Cancer Res 42: 1020

137. Bogacheva TJ, Mirgorodskaya OA, Moskvichev BV (1977) Some physico-chemical properties of modified trypsin. Biokhimiia (Russ) 42: 609
138. Poznansky MJ (1979) In vitro and in vivo activity of soluble cross-linked urikase-albumin polymers: a model for enzyme therapy. Life Sci 24: 153
139. Poznansky MJ, Bhardwaj D (1980) α-1,4-glucosidase-albumin polymers: in vitro properties and advantages for enzyme replacement therapy. Canad J Physiol Pharmacol 58: 322
140. Holcenberg JS, Schmer G, Teller DC, Roberts JS (1975) Biological and physical properties of succinylated and glycosylated Acinetobacter glutaminase-asparaginase. J Biol Chem 250: 4165
141. Snyder PD, Wold JRF, Bernlohr RW, Dullum C (1974) Stabilization to heat and protease degradation by complexing with antibody and by chemical modification. Biochim Biophys Acta 350: 432
142. Bartholeyns J, Moore S (1984) Pancreatic ribunuclease: enzymic and physiological properties of a cross-linked dimer. Science 186: 444
143. Bartholeyns J, Baudhuin P (1976) Inhibition of tumor cell proliferation by dimerized ribonuclease Proc Natl Acad Sci USA 73: 573
144. Kooistra T, Duursma A, Bouma JHW, Gruber M (1977) Endocytosis and breakdown of proteins by sinusoidal liver cells. Acta Biol Med Germ 36: 1763
145. Winkelhake JL (1977) Effects of chemical modification of antibodies on their clearance from the circulation J Biol Chem 252: 1865
146. Poznansky MJ (1977) Perspectives of soluble cross-linked enzyme polymers for enzyme therapy. In: Chang TMS (ed) Biomedical applications of immobilized enzymes and proteins. Plenum Press, New York, p 341
147. Wiman B, Collen D (1978) Molecular mechanism of physiological fibrinolysis. Nature 272: 549
148. Gaffney PJ, Balkuv-Ulitin SM (eds) (1978) Fibrinolysis: current fundamental and clinical concepts. Academic Press, New York
149. Duckert F (1984) Thrombolytic therapy. Sem in Thromb Hemost 10: 87
150. Sahara AA (1980) The case for fibrinolytic therapy. J Cardiovasc Med 5: 793
151. Chazov EI, Golikov AP (1981) Streptodekaza in acute myocardial infarction. Kardiologia (Russ) 21: 10
152. Robbins KC, Summaria L (1976) Plasminogen and plasmin. In: Lorand L (ed) Methods in enzymology. 45B Academic Press, New York, p 257
153. Barlow GH (1976) Urinary and kidney cell plasminogen activator (urokinase). In: Lorand L (ed) Methods in Enzymology. 45B Academic Press, New York, p 239
154. Paoletti K, Sherry S (eds) (1977) Thrombosis and urokinase. Academic Press, New York
155. Rijken DC, Collen D (1981) Purification and characterization of the plasminogen activator secreted by human melanoma cells in culture. J Biol Chem 256: 7035
156. Bachman F, Kruithof IEKO (1984) Tissue plasminogen activator: chemical and physiological aspects. Semin Thromb Hemost 10: 6
157. Castellino FJ, Sodetz JM, Brockway WJ, Siefring GE (1976) Streptokinase. In: Lorand L (ed) Methods in enzymology. 45B Academic Press, New York, p 244
158. Robbins KC, Summaria I, Wohl RC, Bell WR (1983) The human plasmin-derived light (B) chain streptokinase complex: a second-generation thrombolytic agent. Thromb Haemostas (Stuttgart) 50: 787
159. Kanmatsuse K, Kajwara N, Onikura S, Satoh K, Toda Y, Ando T, Satoh K, Seto H, Hatano M (1982) Intracoronary thrombolysis with urokinase during acute myocardial infarction. Ther Arkh (Russ.) 11: 25
160. Torchilin VP, Maksimenko AV, Mazaev AV (1988) Immobilized enzymes for thrombolytic therapy. In: Mosbach K (ed) Immobilized enzymes. (Meth Enzymol) v. 137. Academic Press, Orlando, p 552
161. Ginger LG, Mather AN (1972) Streptokinase chemically bonded to a carbohydrate matrix. US Patent 3 639 213
162. Ginger LG, Mather AN (1976) Methods of dissolving blood clots and the like with streptokinase chemically bonded to a carbohydrate matrix US Patent 3 980 772
163. Murakami A, Josidzaki H, Okuti M, Mori (1980) Method of obtaining plasminogen activator. Jap Patent 26878/80
164. Iga Y, Kitamura M, Chikamori K (1979) Urokinase derivative and its production method. Jap Patent 113 488

165. Chazov EI, Smirnov VN, Torchilin VP, Tereshin IM, Moskvichev BV, Grinberg GM, Skuya AZ, Kleiner GI (1984 Dextran derivative of fibrinolysin. US Patent 4 446 316
166. Chazov EI, Smirnov VN, Torchilin VP, Tereshin IM, Moskvichev BV (1984) Polysaccharid-derivat der Streptokinase, Verfahren zu dessen Herstellung und Anwendung. FRG Patent DE 3 032 606
167. Smirnov VN, Voronkov YuI, Torchilin VP, Mazaev AV (1986) Development of biocompatible preparations of immobilized enzymes and clinical results with immobilized streptokinase (streptodekaza). In: Chazov FI, Smirnov VN (eds) Thrombosis and thrombolysis. Consultants Bureau, New York, p 113
168. Maksimenko AV, Torchilin VP, Smirnov VN, Chazov EI (1982) Heat resistant water soluble urokinase derivative US Patent 4 349 630
169. Papisov MI, Maksimenko AV Torchilin VP (1985) The optimization of reaction conditions during enzyme immobilization on soluble carboxyl-containing carriers. Enz Microb Technol 7: 11
170. Takayanagi H, Kitada L, Sawada T, Suzuki A (1977) Stabilization of urokinase. Jap Patent 145 591
171. Sumi Y (1977) Urokinase drug Jap Patent 159 387
172. Vairel EG, Goulay J, Choay J (1978) Purification of urokinase. US Patent 4 106 992
173. Holleman WH, Lee S-G, Hung PP (1982) Modified urokinase extend activity and methods of making. US Patent 4 326 033
174. Taratina TM, Illarionova NG, Moskvichev BV (1985) The study of conformationalproperties and conformational stability of streptokinase and its polymeric derivative. Biokhimia (Russ) 50: 1312
175. Shimizu K, Nakahara T, Kinoshita T (1989) Plasminogen activator derivatives. US Patent 4 495 285
176. Rajagopalan S, Gonias S, Pizzo SV (1985) A nonantigenic covalent streptokinase-polyethylene glycol complex with plasminogen activator function. J Clin Invest 75: 413
177. Koide J (1982) Modified streptokinase and its preparation. Jap Patent 118 789
178. Koide A, Suzuki S, Kobayashi S (1982) Preparation of polyethylene glycol-modified streptokinase with disappearance of binding ability towards anti-serum an retention of activity. FEBS Lett 143: 73
179. Torchilin VP, Voronkov YuI, Mazaev AV (1982) Use of immobilized streptokinase (Streptodekaza) for thrombosis treatment. Ther Arkh (Russ.) 54: 21
180. Chazov EI, Smirnov VN, Suvorova LA, Suvorov AV, Mazaev AV, Voronkov YuI, Torchilin VP (1981) Influence of a new native thrombolytic preparation "Streptodekaza" on fibrinolysis system. Kardiologiya (Russ.) 21: 18
181. Chazov EI, Gundorova RA, Romaschenko AD, Makarova VP, Mazaev AV, Voronkov YuI, Torchilin VP (1982) Immobilized streptokinase (Streptodekaza) in treatment of intravitreal haemorrhage. Vestn Oftal'mol (Russ) 4: 61
182. Williamson J, Forrester JV (1972) Urokinase in the treatment of vitreous haemorrhage. Lancet 2: 488
183. Williamson J, Forrester JV (1973) Treatment of vitreous haemorrhage with urokinase Lancet 1. 888
184. Zlotchevsky PM (1979) Thromboembolism of pulmonary artery. Meditsina, Moscow
185. Gaudeul P, Moisson P, Witchitz S (1981) Le traitement thrombolytique. Paris
186. Murano G, Bick RL (1980) Basic concepts of hemostasis and thrombosis. CRC Press Inc., Boka Raton
187. Torchilin VP, Il'ina EV, Streltsova LA, Smirnov VN, Chazov EI (1978) Enzyme immobilization on heparin. J Biomed Mater Res 12: 585
188. Maksimenko AV, Torchilin VP (1985) Water-soluble urokinase derivatives of combined action. Thromb Res 38: 277
189. Maksimenko AV, Torchilin VP, Smirnov VN (1982) Kinetics of urokinase hydrolysis of low molecular weight substrates. Biokhimia (Russ.) 47: 405
190. Newmark J, Abuchowski A, Murano J (1982) Preparation and properties of adducts of streptokinase-plasmin complex with polyethylene glycol and pluronic polyol F 38. J Appl Biochem 4: 185
190a. Robbins KC, Boreisha IG (1987) A covalent molecular weight ~92000 hybrid plasminogen activator derived from human plasmin fibrin-binding and tissue plasminogen activator catalytic domains. Biochemistry 26: 4661

191. Chang TMS (1972) Artificial Cells. Charles C Thomas Publisher, Springfield
192. Chang TMS (1973) Biomedical application of artificial cells. Biomed Eng 8: 334
193. Chang TMS (1977) Experimental therapy using semipermeable microcapsules containing enzymes and other biologically active materials. In: Chang TMS (ed) Biomedical applications of immobilized enzymes and proteins. v. 1, Plenum Press, New York-London, p 147
194. Nixon JR (ed) (1976) Microencapsulation. Marcel Dekker Inc., New York
195. Gutcho MH (1972) Capsule technology and microencapsulation. Noyes Data Corp., New York, Park Ridge
196. Aisina RB, Kazanskaya NF (1986) Microencapsulation of physiologically active compounds and their use in medicine. In: Berezin IV (ed) The use of immobilized enzymes. VINITI Press, Moscow
197. Klei HE, Sundstrom DW, Shim D (1985) Immobilization of enzymes by microencapsulation. In: Woodward J (ed) Immobilized cells and enzymes. A practical approach. IRL Press Ltd., Oxford, Chap 4: 49
198. Miura Y, Miyamoto K, Fujii T, Tukamatsu N, Okazaki M (1975) Activity of enzyme immobilized by microencapsulation. In: Weetall HH, Suzuki S (eds) Immobilized enzyme technology. Plenum Press, New York, London, p 73.
199. Kazanskaya NF, Aisina RB, Berezin IV (1983) Autocatalytic enzyme system for amplification of light signals. Enzyme microb Technol 5: 209
200. Chang TMS (1964) Semipermeable microcapsules. Science 146: 524
201. Mori T, Tosa T, Shibata I (1973) Enzymatic properties of microcapsules containing asparaginase. Biochim Biophys Acta 321: 653
202. Aisina RB, Kazanskaya NF, Lukasheva EV, Berezin IV (1976) Production and properties of microencapsulated α-chymotrpsin. Biokhimia (Russ) 41: 1656
203. Cousineau J, Chang TMS (1977) Formation of amino acid from urea and ammonia by sequential enzyme reaction using a microencapsulated multi-enzyme system. Biochem Biophys Res Commun 79: 24
204. Ostergaard JCW, Martiny SC (1973) The immobilization of β-galactosidase through encapsulation in water-insoluble microcapsules. Biotechnol Bioeng 15: 561
205. Couvreur P, Tulkens P, Roland M, Trouet A, Speiser P (1977) Nanocapsules: a new type of lysosomotropic carrier. FEBS Lett 84: 323
206. Speiser P (1979) Nanocapsules as an ultrafine parenteral drug delivery system. In: Kondo T (ed) Microencapsulation. New techniques and application. Techno Books, Tokyo, p 245
207. Hasegawa M (1982) Coating composition. Jap Patent 58631
208. Sipos T (1978) Preparation of enteric coated digestive enzyme compositions. US Patent 4 079 125
209. Rambourg P, Levy J, Levy MC (1982) Microencapsulation. III. Preparation of invertase microcapsules. J Pharm Sci 71: 753
210. Bala K, Vasudevan P (1982) pH-Sensitive microcapsules for drug release. J Pharm Sci 71: 960
211. Campbell J, Chang TMS (1977) Immobilized multienzyme systems and coenzyme requirements: perspectives in biomedical applications. In: Chang TMS (ed) Biomedical applications of immobilized enzymes and proteins. Plenum Press, New York-London, v. 2 p 281
212. Gardner DL, Emmerling DC (1977) Stabilized microencapsulated urease. In: Chang TMS (ed) Biomedical applications of immobilized enzymes and proteins. Plenum Press, New York, London, v. 1 p 163
213. Gardner DL, Kjellstrand CM, Hassler CR, Fink DJ, Emmerling DC (1984) An orally administered microcapsule system for treating chronic renal failure patients. Appl Biochem Biotechnol 10: 27
214. Kitajima M, Miyano S, Kondo A (1969) Studies on enzyme-containing microcapsules. J Chem Soc Japan, Ind Chem Soc 72: 493
215. May SW, Li NN (1972) The immobilization of urease using liquid-surfactant membranes. Biochem Biophys Res Commun 47: 5
216. Sundaram PV (1973) The kinetic properties of microencapsulated urease. Biochim Biophys Acta 293: 1
217. Sparks RE, Salemme RM, Meier RM, Litt MH, Lindan O (1969) Removal of waste metabolites in uremia by microencapsulated reactants. Trans Am Soc Artif Intern Organs 15: 353
218. Poznansky MJ, Chang TMS (1974) Comparison of enzyme kinetics and immunological properties of catalase immobilized by microencapsulation and catalase in free solution for enzyme replacement. Biochim Biophys Acta 334: 103

219. Chang TMS, Poznansky MJ (1968) Semipermeable microcapsules containing catalase for enzyme replacement in acatalasaemic mice. Nature (London) 218: 243
220. Chang TMS (1969) Asparaginase-loaded semipermeable microcapsules for mouse lymphomas. Proc Can Fed Biol Sci 12: 62
221. Chang TMS (1973) Semipermeable microcapsules containing asparaginase: in-vitro and in-vivo stability. Enzyme 14: 95
222. Chong ES, Chang TMS (1974) In vivo effects of intraperitoneally injected L-asparaginase solution and L-asparaginase immobilized within semipermeable nylon microcapsules with emphasis on blood L-asparaginase, "body" L-asparaginase, and plasma L-asparagine levels. Enzyme 18: 218
223. Chang TMS (1971) The in vivo effects of semipermeable microcapsules containing L-asparaginase on 6C3 HED lymphosarcoma. Nature (London) 229: 117
224. Chang TMS, MacIntosh FC, Mason SG (1966) Semipermeable aqueous microcapsules: preparation and properties. Can J Physiol Pharmacol 44: 115
225. O'Grady P, Patrick J (1981) Microencapsulation of bovine liver arginase: characterization and in vivo evaluation of its effect on the growth of the L 1210 murine leukemia. Enzyme Microb Technol 3: 149
226. Shu CD, Chang TMS (1980) Tyrosinase immobilized within artificial cells for detoxification in liver failure: I. Preparation and in vitro studies. Int J Artif Organs 3: 287
227. Wood DA, Whateley TL, Florence AT (1979) Microencapsulation of histidase for enzyme replacement therapy. J Pharm Pharmnacol 31: 79 P Suppl
228. Terman DS, Tavel A, Tavel T (1976) Degradation of circulating DNA by extracorporeal circulation over nuclease immobilized on nylon microcapsules. J Clin Invest 57: 1201
229. Boguslaski RC, Janik AM (1971) A kinetic study of microencapsulated bovine carbonic anhydrase. Biochim Biophys Acta 250: 260
230. Aisina RB, Gracheva II, Karpukova LP, Kazanskaya NF (1980) Bimolecular protein-protein interactions in microcapsules. Biokhimia (Russ) p 45
231. Lukasheva EV, Aisina RB, Gracheva II, Kazanskaya NF, Berezin IV (1977) The properties of α-chymotrypsin enclosed into polycarbonate microcapsules. Estimation of the diffusion effect. Biokhimia (Russ.) 42: 2013
232. Paine MA, Carbonell RG (1975) Immobilization of β-galactosidase in collodion microcapsules. Biotechnol Bioeng 17: 617
233. Chang TMS (1973) Immobilization of enzymes adsorbents or both within semipermeable microcapsules (artificial cells) for clinical and experimental treatments of metabolite-related disorders. In: Bergsma D (ed) Enzyme therapy for congenital diseases. Williams and Wilkins, Baltimore, p 66
234. Kitajima M, Yamaguchi T, Kondo A, Muroya N (1972) Encapsulation method. US Patent 3691090
235. Lourie A, Flechner W (1973) Procede de preparation de microcapsules. France Patent 2169316
236. Bourget L, Chang TMS (1985) Phenylalanine ammonia-lyase immobilized in semipermeable microcapsules for enzyme replacement in phenylketonuria. FEBS Lett 180: 5
237. Bourget L, Chang TMS (1986) Phenylalanine ammonia-lyase immobilized in microcapsules for the depletion of phenylalanine in plasma in phenylketonuric rat model. Biochim Biophys Acta 883: 432
238. Chang TMS (1972) Effects of local applications of microencapsulated catalase on the response of oral lesions to hydrogen peroxide in acatalasemia. J Dent Res 51: 319
239. Levy MC, Rambourg P, Levy J, Potron G (1982) Microencapsulation IV: Cross-Linked hamoglobin microcapsules. J Pharm Sci 71: 759
240. Arakawa M, Kondo T (1980) Preparation and properties of poly (N—α—N—ε—L-lysinediylterephthaloyl) microcapsules containing hemolysate in the nanometer range. Can J Physiol Pharmacol 58: 183
241. Woodland JHR, Yolles S, Blake DA (1973) Long-acting delivery systems for narcotic antagonists. J Med Chem 16: 897
242. Ties C (1979) Properties of injectable microcapsules containing pharmaceuticals. In: Kondo T (ed) Microencapsulation. New techniques and application. Techno Books Tokyo p 143
243. Wakiyama N, Juni K, Nakano M (1982) Preparation and evaluation in vitro and in vivo of polylactic acid microspheres containing dibucaine. Chem Pharm Bull 30: 3719
244. Manenkova MA, Aisina RB, Sorokina NV (1983) Biodegradable microcapsules made of d,l-polylactic acid. Proceedings of IV USSR symposium on enzyme engineering. Kiev, v. 11: 112

245. Maulding HV, Tice TR, Cowsar DR, Fong JW, Pearson JE, Nazareno JP (1986) Biodegradable microcapsules: acceleration of polymeric exipient hydrolytic rate by incorporation of a basic medicament. J Contr Release 3: 103

245a. Lung C, Sklenar G, Ristau O, Pommerening K, Rein H (1987) p-Nitrophenetole deethylase activity of rat liver microsomes entrapped in polyelectrolyte capsules. J Microencapsulation 4: 1

246. Lim F, Sun AM (1980) Microencapsulated islets as bioartificial endocrine pancreas. Science 210: 908

247. Goosen MFA, O'Shea GM, Gharapetian HM, Chou S (1985) Optimization of microencapsulation parameters: semipermeable microcapsules as a bioartificial pancreas. Biotechnol Bioeng 27: 146

248. Kondo T (ed) (1979) Microencapsulation. New techniques and application. Techno Books, Tokyo

249. Chang TMS (1984) Artificial cells in medicine and biotechnology. Appl Biochem Biotechnol 10: 5

250. Widder KJ, Green R (1985) Drug and enzyme targeting. Meth Enzymol 112 part A section I Academic Press, Orlando

251. Chang TMS, Yu YT, Grunwald J (1982) Artificial cell immobilized multienzyme systems and cofactors. In: Chibata I, Fukui S, Wingard Jr LB (eds) Enzyme engineering, v. 6. Plenum Press, New York, p 451

252. Chang TMS, Shu CD, Yu YT, Grunwald J (1982) Artificial cell immobilized enzymes for metabolic disorders. In: Crawfurt M, Gibbs D, Watts RWE (eds) Advanced in the treatment of inborn errors of methabolism. John Wiley and Sons Ltd, UK, p 175

253. Papahadjopoulos D (ed) (1978) Liposomes and their use in biology and medicine. Ann NY Acad Sci, New York

254. Gregoriadis G, Allison AC (eds) (1980) Liposomes in biological systems. John Wiley and Sons, New York

255. Knight CG (ed) (1981) Liposomes: From physical structure to therapeutic applications. Elsevier/North Holland, New York

256. Gregoriadis G (ed) (1984) Liposome Technology vol. 1–3. CRC Press, Boca Raton

257. Schmidt KH (ed) (1986) Liposomes as Drug Carriers. Georg Thieme Verlag, Stuttgart

258. Connor J, Huang L (1986) pH-sensitive immunoliposomes as an efficient and target-specific carrier for antitumor drugs. Cancer Res 46: 3431

259. Desnick RJ, Fiddler MB, Thorpe SR, Steger LD (1977) Enzyme entrapment in erythrocytes and liposomes for the treatment of lysosomal storage diseases. In: Chang TMS (ed) Biomedical applications of immobilized enzymes and proteins. Plenum Press, New York, p 227

260. Szoka F, Papahadjopoulos D (1980) Comparative properties and methods of preparation of lipid vesicles (liposomes). Ann Rev Biophys Bioeng 9: 467

261. Philippot JR, Mutaftschiev S, Liautard JP (1985) Extemporaneous preparation of large unilamellar liposomes. Biochim Biophys Acta 821: 79

262. Shew RL, Deamer DW (1985) A novel method for encapsulation of macromolecules in liposomes. Biochim Biophys Acta 816: 1

263. Madden TD, Bally MB, Hope MJ, Cullis PR, Schieren HP, Janoff AS (1985) Protection of large unilamellar vesicles by trehalose during dehydration: retention of vesicle contents. Biochim Biophys Acta 817: 67

264. Crowe LM, Crowe JH, Rudolph A, Womersley C, Appel L (1985) Preservation of freeze-dried liposomes by trehalose. Arch Biochem Biophys 242: 240

265. Ivanov NN, Rykov SV, Isakova OL, Ruuge EK, Torchilin VP (1985) Estimation of liposome integrity by H-NMR-spectroscopy. Anal Biochem 147: 280

266. Mc Dougall IR, Dunnick J, Mc Namee MG, Kriss JP (1974) Distribution and fate of synthetic lipid vesicle in the mouse. A combined radionuclide and spin label study. Proc Natl Acad Sci USA, 71: 3487

267. Mc Dougall IR, Dunnick JK, Goris ML, Kriss JP (1975) In vivo distribution of vesicles loaded with radiopharmaceuticals: a study of different routes of administration. J Nucl Med 16: 488

268. Abra RM, Hunt CA (1982) Liposome disposition in vivo. IV. The interaction of sequential doses of liposomes having different diameters. Rev Commun Chem Pathol Pharmacol 36: 17

269. Yotsuyanagi T (1983) Therapeutic carrier aspects of orally administered liposomes. Ann Rept Pharm Nagoya City Univ 31: 1

270. Woodley JF (1985) Liposomes for oral administration of drugs. CRC Crit Rev Ther Drug Carrier Syst 2: 1

271. Finkelstein M, Weissmann G (1978) The introduction of enzymes into cells by means of liposomes. J Lip Res 19: 289
272. Weissmann G, Finkelstein M (1980) Uptake of enzyme-bearing liposomes by cells in vivo and in vitro. In: Gregoriadis G, Allison AC (eds) Liposomes in biological systems. John Wiley and Sons, New York, p 153
273. Scherphof G, Roerdink F, Hoekstra D, Zborowksi J, Wisse E (1980) Stability of liposomes in presence of blood constituents: consequences of uptake of liposomal lipid and entrapped compounds by rat liver cells. In: Gregoriadis G, Allison AC (eds) Liposomes in biological systems. John Wiley and Sons, New York, p 179
274. Scherphof G, Damen J, Hoekstra D (1981) Interaction of liposomes with plasma proteins and components of the immune system. In: Knight CG (ed) Liposomes: From physical structure to therapeutic applications. Elsevier/North Holland, New York, chap 10
275. Torchilin VP, Ivanov NN, Klibanov AL, Papisov MI, Chebanov SM (1988) On the mechanism of electron-dense liposomes internalization by macrophages in vitro. In: Gregoriadis G (ed) Liposomes as drug carriers. John Wiley and Sons, Chichester, p 63
276. Roerdink FH, Dijkstra J, Spanjer HH. Scherphof GL (1984) Interaction of liposomes with hepatocytes and Kupffer cells in vivo and in vitro. Bioch Soc Trans 12: 335
277. Senior J, Gregoriadis G (1984) Role of lipoproteins in stability and clearance of liposomes administered to micl. Bioch Soc Trans 12: 339
278. Chebanov SM, Torchilin VP, Berdichevsky VR, Loginov AS, Smirnov VN (1982) The modelling of cholesterol accumulation process in the microcirculatory bed of murine liver with the use of liposomes. Doklady AN SSSR (Russ.) 267: 497
279. Chebanov SM, Torchilin VP, Berdichevsky VR, Loginov AS, Smirnov VN (1983) Ultrastructure of myelinlike particles in the liver of mice administered with liposomes of varying content of cholesterol and phosphatidylcholine. Bull Exp Biol Med (Russ.) 45(4) 110
280. Sessa G, Weissmann G (1970) Incorporation of lysozyme into liposomes. A model for structure-linked latency. J Biol Chem 245: 295
281. Dapergolas G, Neerunjun ED, Gregoriadis G (1976) Penetration of target areas in the rat by liposome-associated bleomycin, glucose oxidase and insulin. FEBS Lett 63: 235
282. Kataoka T, Williamson JR, Kinsky SC (1973) Release of macromolecular markers (enzymes) from liposomes treated with antibody and complement. Attempt at correlation with electron microscopic observations. Biochim Biophys Acta 298: 158
283. Reynolds GD, Barker HJ, Reynolds RH (1978) Enzyme replacement using liposome carriers in feline GM gangliosidosis fibroblasts. Nature 275: 754
284. Steger LD, Desnick RJ (1977) Enzyme therapy. VI. Comparative in vivo fates and effects on lysosomal integrity of enzyme entrapped in negatively and positively charged liposomes. Biochem Biophys Acta 464: 530
285. Braidman I, Gregoriadis G (1976) Preparation of glucocerebroside β-glucosidase for entrapment in liposomes and treatment of patient with adult Gauche's disease. Bioch Soc Trans 4: 259
286. Patel HM, Ryman BE (1974) α-Mannosidase in zinc-deficient rats. Possibility of liposomal therapy in mannosidosis. Bioch Soc Trans 2: 1014
287. Gregoriadis G, Ryman BE (1972) Fate of protein-containing liposomes injected into rats. An approach to the treatment of storage diseases. Eur J Biochem 24: 485
288. Tyrrell DA, Ryman BE, Keeton BR, Dubowitz V (1976) Use of liposomes in treating type II glycogenosis. Br Med J 2: 88
289. Cohen CM, Weissmann G, Hoffstein S, Awasthi Y, Srivastava SK (1976) Introduction of purified hexoseaminidase A into Tay-Sachs leucocytes by means of immunoglobulin-coated liposomes. Biochemistry 15: 452
290. Weissmann G, Bloomgraden D, Kaplan R, Cohen C, Hoffstein S, Collins I, Gottieb A, Nagle D (1975) A general method for the introduction of enzymes by means of immunoglobulin-coated liposomes into lysosomes of deficient cells. Proc Natl Acad Sci, USA, 72: 88
291. Wisse E, Gregoriadis G (1975) The uptake of liposomes by the rat liver. Res J Reticuloendothelial Soc 18: 10a
292. Magee W, Goff SW, Schoknecht J, Smith MD, Cherian K (1974) The interaction of cationic liposomes containing entrapped horseraddish peroxidase with cells in culture. J Cell Biol 63: 492
293. Gregoriadis G, Ryman BE (1972) Lysosomal localization of β-fructofuranosidase-containing liposomes injected into rats Biochem J 129: 123
294. Gregoriadis G, Buckland RA (1973) Enzyme-containing liposomes alleviate a model for storage diseases. Nature 244: 170

295. Gregoriadis G, Neerunjun DE (1974) Control of the rate of hepatic uptake and catabolism of liposomes-entrapped proteins injected into rats: possible therapeutic applications. Eur J Biochem 47: 179
296. Gregoriadis G, Putnam D, Louis L, Neerunjun D (1974) Comparative effect and fate of non-entrapped and liposome-entrapped neuraminidase injected into rats. Biochem J 140: 323
297. Almeida JD, Brand CM, Edwards DC, Heath TD (1975) Formation of virosomes from influenza subunits and liposomes. Lancet 2: 899
298. Turrens JF, Crapo JD, Freeman BA (1984) Protection against oxygen toxicity by intravenous injection of liposome-entrapped catalase and superoxide dismutase. J Clin Invest 73: 87
299. Neerunjun DE, Gregoriadis G (1976) Tumor regression with liposome-entrapped asparaginase: some immunological advantages. Bioch Soc Trans 4: 133
300. Fishman Y, Citri N (1975) L-asparaginase entrapped in liposomes: preparation and properties. FEBS Lett 60: 17
301. Eytan G, Matheson MJ, Racker E (1975) Incorporation of biologically active proteins into liposomes. FEBS Lett 57: 121
302. Redwood WR, Patel BC (1974) Binding of a solubilized membrane ATPase to phospholipid bilayers. Biochim Biophys Acta 363: 70
303. Colley CM, Ryman BE (1974) A model for lysosomal storage diseases and a possible method of therapy. Biochem Soc Trans 2: 871
304. Colley CM, Ryman BE (1976) The use of a liposomally entrapped enzyme in the treatment of an artificial storage condition. Biochim Biophys Acta 451: 417
305. Desnick RJ, Thorpe SR, Fidler MB (1976) Toward enzyme therapy for lysosomal storage diseases. Physiol Rev 56: 57
306. Grabowski GA, Desnick RJ (1981) Enzyme replacement in genetic diseases. In: Holcenberg JS, Roberts J (eds) Enzymes as drugs. John Wiley and Sons, New York
307. Johnson WG, Desnick RJ, Long DM, Sharp HL, Krivit W, Brady B, Brady RO (1973) Intravenous injection of purified hexosaminidase A into a patient with Tay-Sachs diseases. In: Bergsma D (ed) Enzyme therapy in genetic diseases. William and Wilkins, Baltimore, Md vol 9: 120
308. Brady RO, Pentchev P, Gal A, Hibbert S, Dekaban A (1974) Replacement therapy for inherited enzyme deficiency. Use of purified glucocerebrosidase in Gaucher's diesease. N Engl J Med 291: 989
309. Brady RO, Tallman JF, Johnson WG, Gal AE, Leahy WR, Quirk JM, Dekaban AS (1973) Replacement therapy for inherited enzyme deficiency. Use of purified ceramidetrihexosidase in Fabry's disease. N Engl J Med 289: 9
310. Gregoriadis G (1978) Liposomes in the therapy of lysosomal storage diseases. Nature 275: 695
311. Gregoriadis G, Dean MF (1979) Enzyme threrapy in genetic diseases. Nature 278: 603
312. Braidman IP, Gregoriadis G (1977) Rapid partial purification of placental glucocerebroside β-glucosidase and its entrapment in liposomes. Biochem J 164: 439
313. Reynolds GD, Baker HJ, Reynolds RH (1978) Enzyme replacement using liposome carriers in feline G_{M1} gangliosidosis fibroblasts. Nature 275: 754
314. Das PK, Murray GJ, Zirzow GC, Brady RO, Barranger JA (1986) Lectin-specific trageting of β-glucocerebrosidase to different liver cells via glycosylated liposomes. Biochem Med 33: 124
315. Alving CR, Steck EA, Chapman WL Jr, Waits VB, Hendricks LD, Swartz GM Jr, Hanson WL (1978) Therapy of leishmaniasis: superior efficacies of liposome encapsulated drugs. Proc Natl Acad Sci USA 75: 2959
316. New RRC, Chance ML, Thomas SC, Peters W (1978) Antileishmanial activity of antimonial entrapped in liposomes. Nature 272: 55
317. Belchetz PE, Crawley JC, Braidman IP, Gregoriadis G (1977) Treatment of Gaucher's disease with liposome-entrapped glucocerebroside: β-glucosidase. The Lancet 8020: 116
318. Caride VJ, Zaret BL (1977) Liposome accumulation in regions of experimental myocardial infarction. Science 198: 735
319. Bkaily G, Sperelakis N, Elishalom Y, Barenholz Y (1983) Effect of Na^+- or Ca^{2+}-filled liposomes on electrical activity of cultured heart cells. Am J Physiol 245: H756
320. Deshmukh DS, Bear WD, Wisniewski HM, Brocherhoff H (1978) Long-living liposomes as potential drug carriers. Biochem Biophys Res Commun 82: 328
321. Torchilin VP, Berdichevsky VR, Goldmacher VS, Smirnov VN, Chazov EI (1979) Dynamics of blood clearance of liposomes prepared from nonhydrolyzable diether analogue of phosphatidylcholine after intravenous injection into mice. Bull Exper Biol Med (Russ) 8: 160
322. Torchilin VP, Berdichevsky VR, Barsukov AA, Smirnov VN (1980) Coating liposomes with protein decreases their capture by macrophages. FEBS Lett 111: 184

323. Torchilin VP, Khaw BA, Berdichevsky VR, Barsukov AA, Klibanov AL, Haber E, Smirnov VN (1983) Complexes of liposomes with immunoglobulins and sialoglycoproteins. Bull Exp Biol Med (Russ) 46: 51
324. Regen SL, Singh A, Oehme G, Singh M (1981) Polymerized phosphatidyl choline vesicles. Stabilized and controllable time-release carriers. Biochem Biophys Res Commun 101: 131
325. Fidler IJ (1980) Therapy of spontaneous metastases by intravenous injection of liposomes containing lymphokines. Science 208: 1469
326. Kleinerman ES, Fogler WE, Fidler IJ (1985) Intracellular activation of human and rodent macrophages by human lymphokines encapsulated in liposomes. J Leucocyte Biol 37: 571
327. Freeman BA, Young SL, Crapo JD (1983) Liposome-mediated augmentation of superoxide dismutase in endothelial cells prevents oxygen injury. J Biol Chem 258: 12534
328. Freeman BA, Turrens JF, Mirza Z, Crapo JD, Young SL (1985) Modulation of oxidant lung injury by using liposome-entrapped superoxide dismutase and catalase. Fed Proc 44: 2591
328a. Yagi K, Naoi M, Sakai H, Abe H, Konishi H, Arichi S (1982) Incorporation of enzyme into the brain by means of liposomes of novel composition. J Appl Biochem 4: 121
329. Naoi M, Yagi K (1980) Incorporation of enzyme through blood-brain barrier into the brain by means of liposomes. Biochem Intl 1: 591
330. Tokes ZA, St Peteri AK, Todd JA (1980) Availability of liposome content to the nervous system. Liposomes and the blood-brain barrier. Brain Res 188: 292
331. Mokeran R, Firth G, Oliver S, Uttley D, O'Laoire S (1985) A potential application for the intracerebral injection of drugs entrapped within liposomes in the treatment of human cerebral gliomas. J Neurol Neurosurg Psych 48: 1213
332. Osanai T, Nagai Y (1984) Suppression of experimental allergic encephalomyelitis with liposome-encapsulated protease inhibitor: therapy through the blood-brain barrier. Neurochem Res 9: 1407
333. Megaw J (1983) Delivery of liposomally encapsulated drugs to the ocular lens. Res 1: 221
334. Megaw J, Takei Y, Lerman S (1981) Lectin-mediated binding of liposomes to the ocular lens. Exp Eye Res 32: 395
335. Megaw J, Gardner K, Lerman S (1981) Intraocular liposome drug delivery system. Invest Ophthalmol Vis Sci 20(Suppl.) 66
336. Umezawa F, Eto Y, Tokoro T, Ito F, Maekawa K (1985) Enzyme replacement with liposomes containing beta-galactosidase from Charonia lumpas in murine globoil cell leukodystrophy (Twitcher). Biochem Biophys Res Commun 127: 663
337. Espinola LG, Wider EA, Stella AM, Battle AMDC (1983) Enzyme replacement therapy in porphyrias—II: Entrapment of δ-aminolaevulinate dehydratase in liposomes. Int J Biochem 15: 439
338. Radany EW, Bellet RA, Garbers DL (1985) The incorporation of a purified membrane-bound form of guanylate cyclase into phospholipid vesicles and erythrocytes. Biochim Biophys Acta 812: 695
339. Szebeni J, Di Iorio EE, Hauser H, Winterhalter KH (1985) Encapsulation of hemoglobin in phospholipid liposomes: characterization and stability. Biochem 24: 2827
340. Kato A, Tanaka I (1985) Liposome-type artificial red blood cells stabilized with carboxymethyl chitin. Biomat Med Dev Art Org 13: 61
341. Tsuchida E, Hasegawa E, Matsushita Y, Eshima K, Yuasa M, Nishide H (1985) Polymerized liposome as the carrier of heme. A physically stable oxygen carrier under physiological conditions. Chem Lett p 969
342. Hayward JA, Levine DM, Neufeld L, Simon SR, Johnston DS, Chapman D (1985) Polymerized liposomes as stable oxygen-carriers. FEBS Lett 187: 261
343. Kirby CJ, Gregoriadis G (1984) Preparation of liposomes containing factor VIII for oral treatment of haemophilia. J Microencaps 1: 33
344. Papioannou S, Yang P-C, Novotney R (1978) Encapsulation of angiotension II in liposomes: characterization in vitro and in vivo. Clin Exper Hypertension 1: 407
345. Ueno M, Nakasaki T, Horikoshi I, Sakuragawa N (1982) Oral administration of liposomally-entrapped heparin to beagle dogs. Chem Pharm Bull 30: 2245
346. Ueda S, Uchida T, Okada Y (1981) Selective killing of subacute sclerosing panencephalitis virus-infected cells by liposomes containing fragment A of diphtheria toxin. Exper Cell Res 132: 259
347. Rowland RN, Woodley JF (1981) Uptake of free and liposome-entrapped horseradish peroxidase by rat intestinal sacs in vitro. FEBS Lett 123: 41

348. Muller U, Munz K, Waser PG (1983) Incorporation of small unilamellar liposomes loaded with horseradish peroxidase into isolated nerve endings from electric organ of Torpedo marmorata. J Neurocytol 12: 507
349. Viniegra S, Franko J, Cortes E (1984) Incorporation of bovine thyroid peroxidase in liposomes. Int J Biochem 16: 1167
350. Gad AE, Broza R, Eytan GD (1979) Fusion of cells and proteoliposomes: incorporation of beef heart cytochrome oxidase into rabbit erythrocytes. FEBS Lett 102: 230
351. Zakrevsky VI, Plekhanova NG, Khramov VA (1986) Some properties of liposome-encapsulated urease. Ukr Biochem J (Russ.) 58(4): 31
352. Puget K, Michelson AM, Durosay P, Perdereau B, Barbaroux C (1983) Studies on liposomal glutathione transferases. Molec Physiol 3: 43
353. Zajac-Kaye M, Ts'o POP (1984) DNAase I encapsulated in liposomes can induce neoplastic transformation of syrian hamster embryo cells in culture. Cell 39: 427
354. Westcott KR, Hill RL (1985) Reconstitution of a porcine submaxillary gland β-D-galactoside $\alpha 2 \rightarrow 3$ sialyltransferase into liposomes. J Biol Chem 260: 13116
355. Cantenys D, Creach Y, Pradayrol L, Hollande E, Massol M (1983) Covalent attachment of insulin to the outer surface of liposomes. Biochem Biophys Res Commun 117: 399
356. Claassen E, van Rooijen N (1983) A comparative study on the effectiveness of various procedures for attachment of two proteins (L-asparaginase and horse radish peroxidase) to the surface of liposomes. Prepar Biochem 13: 167
357. Vlasov GS, Salov VF, Torchilin VP, Berdichevsky VR (1982) Liposomes and perspectives of their use in applied immunology. Zurn Microbiol Epidemol Immunobiol (Russ) 8(6): 12
358. Hudson LDS, Fiddler MB, Desnick RJ (1979) Enzyme therapy. X. Immune response induced by enzyme and buffer-loaded liposomes in C3H/HeJ Gush mice. J Pharmacol Exp Ther 208: 507
359. Heath TD, Shek P, Papahadjopoulos D (1986) Production of immunogens by antigen cojugation to liposomes. US patent 4 565 696
360. Allen TM, Smuckler EA (1985) Liver pathology accompanying chronic liposome administration in mouse. Res Comm Chem Pathol Pharmacol 50: 281
361. Allen TM, Everest JM (1983) Effect of liposome size and drug release properties on pharmacokinetics fo encapsulated drug in rats. J Pharmacol Exp Ther 226: 539
362. Hunt CA, MacGregor RD (1985) Impact of liposomal drug carriers on drug pharmacokinetics. J Pharm Sci 74: 987
367. Aoyagi J, Yamamoto Y, Akaika T, Nozaki Y (1985) Liposome composition. European patent application 0 160 266
364. Torchilin VP, Klibanov AL, Ivanov NN, Ringsdorf H, Schlarb B (1987) Polymerization of liposome-encapsulated hydrophilic monomers. Makromol Chem Rapid Commun 8: 457
365. Patel HM (1985) Liposomes as a controlled-released system. Biochem Soc Trans 13: 513
366. Petrukhina OO, Ivanov NN, Feldstein MM, Vasil'ev AE, Plate NA, Torchilin VP (1986) The regulation of liposome permeability by polyelectrolytes. J Contr Release 3: 137
367. Ziburdy RP, Magin RZ (1985) Microwave-stimulated drug release from liposomes. Radiat Res 103: 266
368. Ihler GM, Glew RH, Schnure FW (1973) Enzyme loading of erythrocytes Proc Natl Acad Sci USA 70: 2633
369. Zimmermann U (1973) Method of separating ionized substances from an aqueous solution. US Patent 4024054
370. Lewis D, Aplar H (1984) Therapeutic possibilities of drugs encapsulated in erythrocytes. Int J Pharm 22: 137
371. Uptike SJ, Wakamiya RT, Lightfoot EN (1976) Asparaginase entrapped in red blood cells: action and survival. Science 193: 681
372. Ang E, Glew R, Ihler G (1977) Enzyme loading of nucleated chicken erythrocytes. Exp Cell Res 104: 430
373. Kinosita K Jr, Tsong TY (1978) Survival of sucrose-loaded erythrocytes in the circulation. Nature 272: 258
374. DeLoach JR, Wayner GG, Craig TM (1981) Imidocarb dipropionate encapsulation and binding to resealed carrier bovine erythrocytes for potential babesiosis chemotherapy. J Appl Biochem 3: 254
375. Humphreys JD, Ihler, GM (1982) Incapsulation of drugs, enzymes and DNA within human and mouse erythrocytes. In: Bundgaard H, Hansen AB, Kofod H (eds) Optimization of drug delivery, Alfred Benzon Symposium 17. Munksgaard, Copenhagen, p 270

376. Ihler GM (1983) Erythrocyte carriers. Pharmacol Ther 20: 151
377. Zimmermann U (1983) Cellular drug-carrier systems and their possible targeting. In: Goldberg E (ed) Targeted drugs. John Wiley and Sons, New York, chap 8
378. Deloach JR, Sprandel U (eds) (1985) Red blood cells as carriers for drugs. Bibl Haematol 51 S. Karger, Basel
379. Deloach JR (1985) Encapsulation and pharmacokinetics of drugs in bovine and canine carrier erythrocytes. In: Deloach JR, Sprandel U (eds) Red blood cells as carriers for drugs. Bibl Haematol 51 S. Karger, Basel, p 36
380. Seeman PH (1967) Transient holes in the erythrocyte membrans during hypotonic hemolysis and stable holes after lysis by saponin. J Cell Biol 32: 55
381. Lieber MR, Steck TL (1982) A description of the holes in human erythrocyte membrane ghosts. J Biol Chem 257: 11651
382. Deloach JR (1985) Hypotonic dialysis encapsulation in erythrocytes of mammalian species. In: Deloach JR, Sprandel U (eds) Red blood cells as carriers for drugs. Bibl Haematol 51 S. Karger, Basel, p 1
383. Schlegel RA, Rechsteiner MC (1978) Red cell-mediated microinjection of macromolecules into mammalian cells. Methods in cell biology 20: 341
384. Green R (1985) Red cell ghost-entrapped deferoxamine as a model clinical targeted delivery system for iron chelators and other compounds. In: Deloach JR, Sprandel U (eds) Red blood cells as carriers for drugs. Bibl Haematol 51 S Karger, Basel, p 25
385. Chalmelrs D (1985) Comparison and potential of hypo-osmotic and iso-osmotic erythrocytes ghosts and carrier erythrocytes as drug and enzyme carriers. In: Deloach JR, Sprandel U (eds) Red blood cells as cariers for drugs. Bibl Haematol 51 S Karger, Basel, p 15
386. Fiddler MB, Hudson LDS, White J, Desnick RJ (1980) Enzyme therapy. XIV. Comparison of methods for enzyme entrapment in human erythrocytes. J Lab Clin Med 96: 307
387. Deloach JR, Ihler GM (1977) A dialysis procedure for loading erythrocytes with enzymes and lipids. Biochim Biophys Acta, 496: 136
388. Kruse CA, Tin GW, Baldeschwieler JD (1983) Stability of erythrocyte ghosts: A γ-ray perturbed angular correlation study. Proc Natl Acad Sci USA 80: 1212
389. Kinosita K Jr, Tsong TY (1977) Voltage-induced pore formation and hemolysis of human erythrocytes. Biochim Biophys Acta 471: 227
390. Kinosita K Jr, Tsong TY (1977) Formation and resealing of pores of controlled sizes in human erythrocytes membrane. Nature 268: 438
391. Updike SJ (1985) Entrapment of L-asparaginase in red blood cells. In: Deloach JR, Sprandel U (eds) Red blood cells as carriers for drugs. Bibl Haematol 51 S Karger, Basel, p 65
392. Sprandel U (1985) Erythrocytes as carriers for therapeutic enzymes—an approach towards enzyme therapy of inborn errors of metabolism. In: Deloach JR, Sprandel U (eds) Red blood cells as carriers for drugs. Bibl Haematol 51 S Karger, Basel, p 7
393. Way JL, Leung P, Ray L, Sander C (1985) Erythrocyte encapsulated thiosulfate sulfurtransferase. In: Deloach JR, Sprandel U (eds) Red blood cells as carriers for drugs. Bibl Haematol 51 S Karger, Basel, p 75
394. Dale GL, Villacorte DG, Beutler E (1978) High-yield entrapment of proteins into erythrocytes. Biochem Med 18: 220
395. Pitt E, Johnson CM, Lewis DA, Jenner DA, Offord RE (1983) Encapsulation of drugs in intact erythrocytes: an intravenous delivery system. Biochem Pharmacol 32: 3359
396. Alpar HO, Lewis DA (1985) Therapeutic efficacy of asparaginase encapsulated in intact erythrocytes. Biochem Pharmacol 34: 257
397. Mattarella N, Richardson T (1979) Derivatives of L-asparaginase designed to be immobilized in vivo. Bioch Soc Trans 7: 66
398. Dale GL, Beutler E (1976) Enzyme replacement therapy in Gaucher's disease: a rapid high-yield method for purification of glucocerebrosidase. Proc Natl Acad Sci USA 73: 4672
399. Beutler E, Dale GL, Guinto E, Kuhl W (1977) Enzyme replacement therapy in Gaucher's disease: preliminary clinical trial of a new enzyme preparation. Proc Natl Acad Sci USA 74: 4620
400. Sprandel U, Hubbard AR, Chalmers RA (1981) Towards enzyme therapy using erythrocytes. J Inher Metab Dis 4: 99
401. Fiddler MB, Hudson LDS, Desnick RJ (1977) Immunological evaluation of repeated administration of erythrocyte-entrapped protein to C3H/HeJ mice. Biochem J 168: 141

402. Kashkin AP, Levy EV, Putilova NE, Mahovenko LV, Verblovskaya NG, Kuzina ZA, Moskvichev BV (1982) Izuchenie immunnogo otveta k preparatam immobilizovannyh fermentov (Immune response to immobilized enzymes). In: Moskvichev BV (ed) Immobilizovannye fermently v meditsine i meditsinskoy promyshlennosty (Immobilized enzymes in medicine and medical industry Russ.) VNITIAF, Leningrad, p 111
403. Ihler G, Lantzy A, Purpura J, Glew RH (1975) Enzymatic degradation of uric acid by uricase-loaded human erythrocytes. J Clin Invest 56: 595
404. Adriaenssens K, Karcher D, Lowenthal A, Terheggen HG (1976) Use of enzyme-loaded erythrocytes in in vitro correction of arginase-deficient erythrocytes in familial hyperargininemia. Clin Chem 22: 323
405. De Flora A, Morelli A, Benatti U (1985) Entrapment of normal and mutant glucose 6-phosphate dehydrogenase (G6PD) within G6PD deficient erythrocytes. In: Deloach JR, Sprandel U (eds) Red blood cells as carriers for drugs. Bibl Haematol 51 S Karger, Basel, p 50
406. Fiorelli G, Ploner T, Berardi W, Manoussakis C (1985) G6PD encapsulation and metabolic properties of loaded erythrocytes. In: Deloach JR, Sprandel U (eds) Red blood cells as carriers for drugs . Bibl Haematol 51, S Karger, Basel, p 57
407. Matas AJ (1976) Hepatocellular transplantation for metabolic deficiencies: decrease of plasma bilirubin in gunn rats. Science 192: 892
408. Levin RB, Sorokin LV, Beresov TT (1979) Adsorptional immobilization of enzymes on blood cells. Doklady AN SSSR (Russ.) 249: 235
409. Docherty PA, Aronson Jr NN (1982) Injection of proteins into primary rat hepatocytes by erythrocyte-mediated techniques. Biochem Biophys Res Commun 109: 527
410. Hosoi T, Ozawa K, Tsao C-J, Urabe A, Uchida T, Takaku F (1985) Introduction of macromolecules into hemopoietic stem cells with an erythrocyte-ghost-mediated system. Biochem Biophys Res commun 128 193
410a. Knowles SF, Hopgood MF, Ballard FJ (1988) Degradation for horseradish peroxidase after microinjection into mammalian cells. Experimental cell research 174: 266
411. Deloach JR, Sprandel U, Green R (1985) Carrier erythrocytes: a prospectus for the future. In: Deloach JR, Sprandel U (eds) Red blood cells as carriers for drugs. Bibl Haematol 51 S Karger, Basel, p 157
412. Baynall RD (1977) A possible classification of controlled drug release. Biomat Med Dev Art Org 5(4): 355
413. Kato T (1983) Encapsulated drugs in targeted cancer therapy. In: Bruck SD, (ed) Controlled drug delivery. CRC Press, Boca Raton, v. 2 p 189
414. Ivanova TP, Mirgorodskaya OA, Moskvichev BV, Samsonov GV (1976) Stability of α-amylase from Bacillus subtilis reversibly immobilized by gel KMT. Prikl Biokhim Mikrobiol (Russ) 12: 33
415. Ostrowski DH, Dmitrienko LB, Samsonov GV, Leibson LG, Yunev OA (1973) Stability of insulin in a complex with polyelectrolytes to the action of pepsin. Khim Farm Zhurn (Russ) 7: 49
416. Edman P, Sjoholm I (1983) Acrylic microspheres in vivo. VI. Antitumor effect of microparticles with immobilized L-asparaginase against 6C3HED lymphoma. J Pharm Sci 72: 854
417. Chibata J, Tosa T, Mori T (1974) Encapsulation of stable, water insoluble uricase. Jap Patent 55818/74
418. Hinberg I, O'Driscall KF (1975) Preparation and kinetic properties of gel entrapped urate exidase. Biotechnol Bioenerg 17: 1435
419. Langer R, Folkman J (1978) Sustained release of macromolecules from polymers. In: Elias HG (ed) Polymeric delivery systems. Midland macromolecular monograph, series V. Gordon and Breach, New York, p 175
420. Langer R, Folkman L (1976) Polymers for the sustained release of proteins and other macromolecules. Nature 263: 797
421. Rhine W, Hsieh D, Langer R (1980) Polymers for sustained macromolecule release: procedures to fabricate reproducible delivery systems and control release kinetics. J Pharm Sci 69: 265
422. Langer R, Brem H, Tapper D (1981) Biocompatibility of polymeric delivery systems for macromolecules. J Biomed Mater Res 15: 267
423. Hsu TTP, Langer R (1985) Polymers for the controlled release of macromolecules: effect of molecular weight of ethylene-vinylacetate copolymer. J Biomed Mater Res 19: 445
424. Torchilin VP, Tischenko EG, Smirnov VN, Chazov EI (1977) Immobilization of enzymes on slowly soluble carriers. J biomed Mater Res 11: 223

425. Torchilin VP, Bobkova AC, Smirnov VN, Chazov EI (1976) Enzyme immobilization on biocompatible carriers. I. Immobilization of α-chymotrypsin on modified Sephadexes. Bioorg Khim (Russ) 2: 116
476. Torchilin VP (1983) Immobilized enzymes and the use of immobilization principles for drug targeting. In: Goldberg EP (ed) Targeted drugs. John Wiley and Sons, New York, p 127
427. Chazov EI, Mazaev AV, Torchilin VP, Lebedev BS, Il'ina EV, Smirnov VN (1977) Study of biosoluble preparations of immobilized fibrinolysin in experiment. Kardiologiya (Russ.) 17: 139
428. Chazov EI, Mazaev AV, Torchilin VP, Lebedev BS, Il'ina EV, Smirnov VN (1978) Experimental study of biosoluble drugs. Thrombus lysis with biosoluble immobilized fibrinolysin in experiment. Thromb Res 12: 809
429. Torchilin VP, Lebedev BS, Mazaev AV, Kukharchuk VV, Eventov AZ, Cramer AA, Smirnov VN, Chazov EI (1976) Study of solution rate of polysaccharide carriers for prolonging the action of drugs in chronic experiments. Kardiologiya (Russ) 16: 102
430. Smirnov VN Voronkov YuI, Torchilin VP, Mazaev AV (1986) Development of biocompatible preparations of immobilized enzymes and clinical results with immobilized streptokinase (streptodekaza). In: Chazov EI, Smirnov VN (eds) Thrombosis and thrombolyusis. Consulting Bureau, New York, p 113
431. Capet-Antonioni FC, Tamenasse J (1975) Kinetic study of soluble and insoluble urokinase. Canad J Biochem 53: 890
432. Grinenko TV, Tret'yatchenko VG (1983) Immobilization of streptokinase on insoluble carriers and properties of immobilized protein. Ukr Biochem J (Russ) 55: 307
433. Sjoholm I, Edman P (1979) Acrylic microspheres in vivo. I. Distribution and elimination of polyacrylamide microparticles after intravenous and intraperitoneal injection in mouse and rats. J Pharmacol Exper Ther 211: 656
434. Edman P, Sjoholm I, Brunk U (1983) Acrylic microspheres in vivo. VII. Morphological studies on mice and cultured macrophages. J Pharm Sci 72: 658
435. Artursson P, Edman P, Laakso T, Sjoholm I (1984) Characterization of polyacryl starch microparticles as carriers for proteins and drugs. J Pharm Sci 73: 1507
436. Sjoholm I, Edman P (1984) The use of biocompatible microparticles as carriers of enzymes and drugs in vivo. In: Davis SS, Illum L, Mc Vie JG, Tomlinson E (eds) Microspheres and drug therapy. Pharmaceutical, immunological and medical aspects. Elsevier Science Publishers BV, New York, p 245
437. Ekman B, Lofter C, Sjoholm I (1976) Incorporation of macromolecules in microparticles: preparation and characteristics. Biochemistry 15: 5115
438. Edman P, Sjoholm I (1979) Acrylic microspheres in vivo. II. The effect in rat of L-asparaginase given in microparticles of polyacrylamide. J Pharmacol Exper Ther 211: 663
439. Edman P, Ekman B, Sjoholm I (1980) Immobilization of proteins in microspheres of biodegradable polyacryldextran. J Pharm Sci 69: 838
440. Artursson P, Edman P, Sjoholm I (1984) Biodegradable microspheres. I. Duration of action of dextranase entrapped in polyacrylstarch microparticles in vivo. J Pharmacol Exper Ther 231: 705
441. Edman P, Sjoholm I (1982) Treatment of artificially induced storage disease with lysosomotropic microparticles. Life Sci 30: 327
442. Hannibal-Friedrich O, Chun M, Sernetz M (1980) Immobilization of β-galactosidase, albumin and γ-globulin on epoxy-activated acrylic beads. Biotechnol Bioeng 22: 157
443. Schram AW, Hamers MN, Oldenbroek-Haverkamp E, Strijland A, de Jonge A, van den Bergh FAJThM, Tager JM (1978) Properties of immobilized β-galactosidase and effect on ceremide-3 content of plasma from patients with Fabry's disease. Biochim Biophys Acta 527: 456
444. Wingard Jr LB, Tritton TR, Egler K (1985) Cell surface effects of adriamycin and carminomycin immobilized on cross-linked polyvinyl alcohol. Cancer Res. 45: 3529
445. Couvreur P, Kante B, Lenaerts V, Scailteur V, Roland M, Speiser P (1980) Tissue distribution of antitumor drugs associated with polyalkylcyanoacrylate nanoparticles. J Pharm Sci 69: 199
446. Simonescu Cr, Dumitriu S, Bulacovschi V, Popa VI (1979) Bioactive polymers. III. Cellulose derivatives as support for biologically active compounds. J Pol Sci Pol Symp 66: 171
447. Singh M, Vasudevan P, Ray AR, Guha SK (1980) Biosoluble polymers for drug delivery. Makromol Chem 181: 2433
448. Rakhimov MM, Zemlyanskaya NR, Avlyanova RR, Tillyabaeva IS (1979) Immobilization of enzymes on globulin. Biokhimia (Russ) 44: 1347

449. Martodam RR, Wumasi DY, Liener IE, Powers JC, Nishino N, Krejcarek G (1979) Albumin microspheres as carrier of an inhibitor of leukocyte elastase: potential therapeutic agent for emphysema. Proc Natl Acad Sci USA 76: 2128
450. Tenovuo J, Kurkijarvi K (1981) Immobilized lactoperoxidase as a biologically active and stable form of an antimicrobial enzyme. Arch Biol 26: 309
451. De Luca M, Kricka LJ (1989) Coimmobilized multienzymes: an in vitro model for cellular processes. Arch Biochem Biophys 226: 285
452. Martinek K, Siksnis VA, Mozhaev VV, Berezin IV, Smirnov VN, Torchilin VP (1982) Principles of enzyme stabilization. Immobilized in polyacrylamide gel trypsin and chymotrypsin, possessing high catalytic activity at elevated temperatures. DAN SSSR (Russ) 263: 494
453. Mozhaev VV, Siksnis VA, Torchilin VP, Martinek K (1983) Operational stability of copolymerized enzymes at elevated temperatures. Biotechnol Bioeng, p 25
454. Valuev LI, Plate NA, Sinani VA, Chupov VV (1987) The pecularities of immobilization of modified proteins with the use of copolymerization reaction. Biotechnol (Russ) 3(2): 173
455. Martinek K, Siksnis VA, Mozhaev VV, Berezin IV, Smirnov VN, Torchilin VP (1981) Regulation of the catalytic activity of an enzyme, immobilized on polyelectrolyte carrier. DAN SSSR (Russ) 259: 746
456. Plate NA, Valuev LI, Chupov VV (1984) The chemistry of macromonomers basing physiologically active substances and some problems of the synthesis of polymers for biomedical application. Pure Appl Chem 56: 1351
457. Kirpotin DV, Motorin YuA, Mevkh AT, Varfolomeev SD, Orlovsky AF, Gladilin KL (1986) Preparation and properties of prostaglandin-H-synthetase immobilized in insoluble polyelectrolyte complexes. Biokhimia (Russ) 51: 1268
458. Nakagame F, Honda H, Maruyama T, Saito N, Kimoto T (1985) Pharmaceutical composition containing urokinase. European Patent Application 0 143 939
459. Balazs AC, Calef DF, Deutch JM, Siegel RA, Langer R (1985) The role of polymer matrix structure and interparticle interactions in diffusion-limited drug release. Biophys J 47: 97
460. Kusserow BK, Larrow RW, Nichols JE (1973) The surface bonded, covalently crosslinked urokinase synthetic surface. In vitro and chronic in vivo studies. Trans Am Soc Artif Intern Organs 19: 8
461. Sugitachi A, Takagi K, Imaoka S (1978) Immobilization of plasminogen activator, urokinase, on nylon. Thromb Haemost 39: 426
462. Sugitachi A, Kawahara T, Kodama J, Kikkawa Y, Takagi K (1978) Antithrombogenicity of immobilized urokinase and clinical significance. Blut 37: 31
463. Cooney DA, Weetall HH, Long E (1975) Biochemical and pharmacologic properties of L-asparaginase bonded to dacron vascular prostheses. Bioch Pharm 24: 503
464. Jefferies SR, Richards R, Bernath FR, Joyeuse R (1977) Preliminary studies with L-asparaginase bound to implantable bovine collagen heterografts: a potential longterm, sustained dosage, antitumor enzyme therapy system. Biomater Med Devices Artif Organs 5: 337
465. Zubairov DM, Zinkevich OD, Asadullin MG (1986) Fibrinolytically active polymers. In: Chazov EI, Smirnov VN (eds) Thrombosis and thrombolysis. Consulting Bureau, New York, p 179
466. Valuev LI (1988) Macromonomers of physiologically active compounds. VINITI Moscow
467. Plate NA, Chupov VV, Valuev LI, Burdigina IF (1985) On the increase of the action selectivity of proteolytic enzymes covalently immobilized in polymeric hydrogel matrix. DAN SSSR (Russ) 282: 1010
468. Plate NA, Valuev LI, Chupov VV, Burdigina IF (1985) Polymers modified by means of biologically active substances—a new class of hemocompatible materials for contact with blood. Voprosy Med Khim (Russ) 31: 43
469. Valuev LI, Chupov VV, Plate NA (1986) On the regulation of the interaction of synthetic polymers with blood, Makromol Chem Macromol Symp 4: 245
470. Shumakov VI, Chepurov AK (1972) Experimental thrombolysis. Kardiologia (Russ) 2: 88
471. Goldberg EP (ed) (1983) Targeted drugs. John Wiley and Sons, New York
472. Gregoriadis G (1977) Targeting of drugs. Nature (London) 265: 407
473. Shaw IH, Dingle JT (1980) Liposomes as steroid carriers in the intraarticular therapy of rheumatoid arthritis. In: Gregoriadis G, Allison AC (eds) Liposomes in biological systems. John Wiley and Sons, Chichester, England, chap 11

474. Chazov EI, Matveeva LS, Mazaev AV, Sargin KE, Sadovskaya GV, Ruda MYa (1976) Intracoronary administraton of fibrinolysin for the treatment of acute myocardial infarction. Ther Arch (Russ) 48(4): 8
475. McLaughlin CA, Goldberg EP (1983) Local chemo- and immunotherapy by intratumor drug injection: opportunities for polymer-drug compositions. In: Goldberg EP (ed) Targeted drugs. John Wiley and Sons, New York
476. Yatvin MB, Weinstein JN, Dennis WH, Blumenthal R (1978) Design of liposomes for enhanced local release of drugs by hyperthermia. Science 202: 1290
477. Weinstein JN, Magin RL, Yatvin MB, and Saharko DS (1989) Liposomes and local hyperthermia: selective delivery of methotrexate to heated tumors. Science 204: 188
478. Sullivan SM, Huang L (1985) Preparation and characterization of heat-sensitive immunoliposomes. Biochim Biophys Acta 812: 116
479. Yatvin MB, Kreutz W, Horwitz BA, Shinitzky M (1980) pH-sensitive liposomes: possible clinical implications. Science 210: 1253
480. Connor J, Yatvin MB, Huang L (1984) pH-sensitive liposomes: acid-induced liposome fusion. Proc Natl Acad Sci USA 81: 1715
481. Nayar R, Schroit AJ (1985) Generation of pH-sensitive liposomes: use of large unilamellar vesicles containing N-succinyldioleoylphosphatidylethanolamine. Biochem 24: 5967
482. Straubinger RM, Duzgunes N, Papahadjopoulos D (1985) pH-sensitive liposomes mediate cytoplasmic delivery of encapsulated macromolecules. FEBS Lett 179: 148
483. Wingard LB Jr (1983) Immobilized drugs and enzymes in biochemical pharmacology, perspectives and critique. Biochem Pharm 32: 2647
484. Ringsdorf H (1975) Structure and properties of pharmacologically active polymers. J Polym Sci Symp No 51 p 135
485. Pratten MK, Duncan R, Lloyd JB (1978) A comparative study of the effects of polyamino acids and dextran derivatives on pinocytosis in the rat yolk sac. Biochim Biophys Acta 540: 455
486. Gregoriadis G (1981) Targeting of drugs: implications in medicine. Lancet 8240: 241
487. Vitetta ES, Krolick KA, Miyama-Inaba M, Cushley W, Uhr JW (1983) Immunotoxins: a new approach to cancer therapy. Science 219: 644
488. Cumber AJ, Forrester JA, Foxwell BML, Ross WCJ, Thorpe PE (1985) Preparation of antibody-toxin conjugates. Widder KH, Green R (eds) Drug and enzyme targeting. Methods in enzymology. Academic Press, London Part A p 207
489. Faulstich H, Fiume L (1985) Protein conjugates of fungal toxins In: Widder KJ, Green R (eds) Drug and enzyme targeting. Methods in enzymology. Academic Press, London, Part A p 225
490. Everson RB, Hall TC, Wittlif JL (1973) Treatment in vivo of R 3230 AC carcinoma of the rat with estradiol mustarel (NSC-112259) or its molecular components. Cancer Chemother Rep 57: 353
491. Wall ME, Abernathy AS Jr Carroll FI, Taylor DJ (1969) The effects of some steroidal alkylating agents on experimental animal mammary tumor and leukemia systems. J Med Chem 12: 810
492. Fiume L, Busi C, Mattioli A (1983) Targeting of antiviral drugs by coupling with protein carriers. FEBS Lett 153: 6
493. Ringsdorf H (1978) Synthetic polymeric drugs. In: Dostelnila R (ed) Polymeric delivery systems. Gordon and Breach, New York, p 197
494. Goldberg EP (1978) Polymeric affinity drugs for cardiovascular, cancer and urolithiasis therapy. In: Donamura LG, Vogel O (eds) Polymeric drugs. Academic Press, New York, p 239
495. Goldberg EP (1978) Polymeric affinity drugs. In: Elias HG (ed) Midland macromolecular monograph, Ser 5. Gordon and Breach, New York, p 227
496. Rihova B, Kopecek J, Kopeckova-Rejmanova P, Strohalm J, Ploccova D (1986) Bioaffinity therapy with antibodies and drugs bound to soluble synthetic polymers. J Chromatogr.—Biomed Appl 276: 211
497. Arnon R, Hurwitz E (1983) Antibody- and polymer-drug conjugates. In: Goldberg EP (ed) Targeted drugs. John Wiley and Sons, New York, chap 2
498. Tsukada Y, Hurwitz E, Kashi R, Sela M, Hibi N, Hara A, Hirai H (1982) Chemotherapy by intravenous administration of cojugates of daunomycin with monoclonal and conventional anti-rat -fetoprotein antibodies. Proc Natl Acad Sci USA 79: 7896
499. Dunkan R, Kopecek J, Rejmanova P, Lloyd JB (1983) Targeting of N-(2-hydroxypropyl)-methacrylamide copolymers to liver by incorporation of galactose residues. Biochim Biophys Acta 755: 518

500. Chang TMS (1976) Microencapsulation of enzymes and biologicals. In: Mosbach K (ed) Immobilized enzymes, methods in enzymology, Vol 44. Academic Press, New York, chap 16
501. Ihler GM (1979) Potential use of erythrocytes as carriers for enzymes and drugs. In: Gregoriadis G (ed) Drug carriers in biology and medicine. Academic Press, New York, p 129
502. Kulkarni PN, Blair AH, Ghose TI (1981) Covalent binding of methotrexate to immunoglobulins and the effect of antibody-linked drug on tumor growth in vivo. Cancer Res 41: 2700
503. Gallego J, Price MR, Baldwin RW (1984) Preparation of four daunomycin-monoclonal antibody 791T/36 conjugates with anti-tumor activity. Int J Cancer 33: 737
504. Tsukada Y, Bischof WK-D, Hibi N, Hirai H, Hurwitz E, Sela M. Effect of a conjugate of daunomycin and antibodies to rat α-fetoprotein on the growth of α-fetoprotein-producin tumor cells. Proc Natl Acad Sci USA, 79: 621
505. Manabe Y, Tsubota T, Haruta Y, Kataoka K, Okazaki M, Haisa S, Nakamura K, Kimura I (1984) Production of a monoclonal antibody-methotrexate conjugate utilizing dextran T-40 and its biological activity. J Lab Clin Med 104: 445
506. Jung G, Kohnlein W, Luders G (1981) Biological activity of the antitumor protein neocarzinostatin coupled to a monoclonal antibody by N-succinimidyl 3-(2-pyridyldithio)-propionate. Biochem Biophys Res Commun 101: 599
507 .Torchilin VP, Maksimenko AV, Tischenko EG, Ignashenkova GV, Ermolin GA (1984) Immobilized thrombolytic enzymes possessing increased affinity towards substrate. In: Laskin AI, Tsao GT, Wingard LB Jr (eds) Enzyme engineering—7. Ann NY Acad Sci 434: 289
508. Torchilin VP, Maksimenko AV, Ignashenkova GV, Tischenko EG, Ermolin GA, Smirnov VN (1984) Fibrinolytic action of an enzymatic preparation conjugated with specific antibodies. Bull Exper Med (Russ) 98: 556
509. Sevilla CL, Mahle NH, Boylan CM, Callewaert DM (1983) Plasminogen activator-anti-human fibrinogen conjugate. Biochem Biophys Res Commun 130: 91
510. Hui KY, Haber E, Matsueda GR (1983) Monoclonal antibodies to a synthetic fibrin-like peptide bind to human fibrin but not fibrinogen. Science 222: 1129
510a. Hui KY, Haber E, Matsueda GR (1985) Immunodetection of human fibrin using monoclonal antibody-6405 in an extracorporeal chiken model. Thromb Haemost 54: 524
511. Bode C, Matsueda GR, Hui KY, Haber E (1985) Antibody-directed urokinase: a specific fibrinolytic agent. Science 229: 765
511a. Bode C, Runge MS, Newell JB, Matsueda GR, Haber Ē (1987) Characterization of an antibody-urokinase conjugate. A plasminogen activator targeted to fibrin. J Biol Chem 262: 10819
512. Runge MS, Bode C, Matsueda GR, Haber E. Antibody-enhanced thrombolysis: trageting of tissue plasminogen activator in vivo. In press
512a. Schnee JM, Runge MS, Matsueda GR, Hudson NW, Seidman JG, Haber E, Quertermous T. Construction and expression of a recombinant antibody-targeted plasminogen activator, in press
512b. Runge MS, Bode C, Matsueda GR, Haber E (1987) Antibody-enhanced thrombolysis: Targeting of tissue plasminogen activator in vivo. Proc Natl Acad Sci USA 84: 7659
513. Maksimenko AV, Torchilin VP (1985) Water-soluble urokinase derivatives with increased affinity to the fibrin clot. Thromb Res 38: 289
514. Maksimenko AV, Torchilin VP (1985) Water-soluble modified derivatives of urokinase: fibrinilytic activity and other properties. Vopr Med Khim (Russ) 31(4): 12
515. Nakayama Y, Miyazaki W, Shinohara M (1985) Urokinase-plasmin complex adsorbable by fibrin and process for preparating same. US Patent 4545988
516. Nakayama Y, Shinohara M, Tani T, Kawaguchi T, Furuta T, Izawa T, Kaise H, Miyazaki W, Nakano Y (1986) The plasmin heavy chain-urokinase conjugate: a specific thrombolytic agent. Thromb Haemost 56: 364
517. Maksimenko AV, Rusetsky AN, Torchilin VP (1987) Fibrinolytic action of an enzyme covalently bound with modified thrombin. Bull Exper Biol Med (Russ) 103: 35
518. Murrayi GJ, Doebber TW, Shen IY, Wu MS, Ponpipom MM, Bugianesi RL, Brady RO, Barranger JA (1985) Targeting of synthetically glycosylated human placental glucocerebrosidase. Biochem Med 34: 241
519. Samokhin GP, Smirnov MD, Muzykantov VR, Domogatsky SP, Smirnov VN (1983) Red blood cells targeting to collagen-coated surfaces. FEBS Lett 154: 257

520. Smirnov MD, Samokhin GP, Muzykantov VR, Idelson GL, Domogatsky SP, Smirnov VN (1983) Type I and III collagens as a possible target for drug delivery to the injured sites of vascular bed. Biochem Biophys Res Commun 116: 99
521. Gregoriadis G (1980) The liposomes drug-carrier concept: its development and future. In: Gregoriadis G, Allison AC (eds) Liposomes in biological systems. John Wiley and Sons, New York, chap 2
522. Weinstein JN (1981) Liposomes as "targeted" drug carriers: a physical chemical perspective. Pure Appl Chem 53: 2242
523. Torchilin VP, Smirnov VN, Chazov EI (1982) Problems and perspectives in liposomes use for drug targeting. Vopr Med Khim (Russ) 28(1): 3
524. Finkelstein MC, Weissmann G (1981) Targeting of liposomes. In: Knight CG (ed) Liposomes: From physical structure to therapeutic applications. Elsevier/North-Halland, New York, chap 16
525. Torchilin VP (1984) Immobilization of specific proteins on liposomes surface: systems for drug targeting. In: Gregoriadis G (ed) Liposome technology, Vol. 3. CRC Press, Boca Raton, Fla, chap 6
526. Shinozawa S, Tsutsui K, Oda T (1979) Enhancement of the antitumor effect of illudin S including it into liposomes. Experientia 35: 1102
527. Shinozawa S, Araki Y, Oda T (1981) Tissue distribution and antitumor effect of liposome-entrapped doxorubicin (adriamycin) in Ehrlich solid tumor-bearing mouse. Acta Med Okayama 35: 395
528. Rosenberg OA, Bekreneva V Yu, Loshakova LV, Rezvaya SP, Davidenkova EF, Hanson KP (1983) Enhanced trapping by tumor cells of liposomes prepared from autologous phospholipids. Vopr Onkol (Russ) 29(10): 56
529. Rozenberg OA, Bekreneva V Yu, Loshakova LV, Rezvaya SP, Davidenkova EF, Khanson KP (1984) Specificity of liposomal uptake from lipids of target cells. Bull Exper Biol Med (Russ) 97(6): 670
530. Burkhanov S, Berdichevsky VR, Saatov TS, Torchilin VP. Increased affinity of liposomes prepared from autologous phospholipids toward the target organ. Biol Membranes (Russ) in press
531. Kayawake S, Kako KJ (1982) Association of liposomes with the isolated perfused rabbit heart. Basic Res Cardiol 77: 668
532. Torchilin VP, Klibanov AL (1981) Immobilization of proteins on liposome surface. Enzyme Microb Technol 3: 297
533. Urdal DL, Hakomori S (1980) Tumor-associated ganglio-N-triosyl ceramide, target for antibody-dependent, avidinmediated drug killing of tumor cells. J Biol Chem 225: 10509
534. Bayer BA, Rivnay B, Skutelsky E (1979) On the mode of liposome-cell interaction: biotin-conjugated lipids as ultrastructural probes. Biochim Biophys Acta 550: 463
535. Leserman LD, Barbet J, Kourilsky F, Weinstein JN (1980) Targeting to cells of fluorescent liposomes covalently coupled with monoclonal antibody or protein A. Nature (London) 288: 602
536. Gregoriadis G, Meehan A, Mah MM (1981) Interaction of antibody-bearing small – unilamellar vesicles with target free antigen in vitro and in vivo. Biochem J 200: 202
537. Kagawa Y, Racker E (1971) Partial resolution of enzyme catalysing oxidative phosphorylation. XXV. Reconstitution of vesicles catalysing P^{32}-adenosine triphosphate exchange. J Biol Chem 246: 5477
538. Hinkle PC, Kim JJ, Racker E (1971) Ion transport and respiratory control in vesicles formed from cytochrome-oxidase and phospholipids. J Biol Chem 247: 1338
539. Racker E (1972) Reconstitution of a calcium pump with phospholipids and a purified Ca^{2+} adenosine triphosphatase from sarcoplasmic reticulum. J Biol Chem 247: 8198
540. Huang L, Kennell SL (1979) Binding of immunoglobulin G to phospholipid vesicles by sonication. Biochemistry 18: 1702
541. Neerunjun ED, Hunt R, Gregoriadis G (1977) Fate of liposome-associated agent injected into normal and tumorbearing rodents: attempts to improve localizations in tumor tissues. Biochem Soc Trans 5: 1380
542. Dunnick JK, McDougall IR, Aragon S, Goris ML, Kriss TP (1975) Vesicle interactions with polyaminoacids and antibody: in vitro and in vivo studies. J Nucl Med 16: 483
543. Endoh H, Suzuki Y, Hashimoto Y (1981) Antibody coating of liposomes with 1-ethyl-3-(3-dimethylaminopropyl)carbodiimide and the effect on target specificity. J Immunol Methods 44: 79

544. Torchilin VP, Goldmacher VS, Smirnov VN (1978) Comparative studies on covalent and noncovalent immobilization of protein molecules on the surface of liposomes. Biochem Biophys Res Commun 85: 983
545. Torchilin VP, Khaw BA, Smirnov VN, Haber E (1979) Preservation of antimyosin antibody activity after covalent coupling to liposomes. Biochem Biophys Res Commun 89: 1114
546. Torchilin VP, Khaw BA, Berdichevsky VR, Locke ER, Smirnov VN, Haber E, Chazov EI (1979) Preservation of antibody activity after covalent coupling to liposomes. DAN SSSR (Russ) 246: 746
547. Budker VG, Mustaev AA, Pressman EK, Roschke VV, Vakhrusheva TA (1982) Adsorption of non-membrane proteins on the surface of model phospholipid membranes. Biochim Biophys Acta 688: 541
548. Heath TD, Robertson D, Birbeck MSC, Davies AJS (1980) Covalent attachment of horseradish peroxidase to the outer surface of liposomes. Biochim Biophys Acta 599: 42
549. Chua M-M, Fan S-T, Karush F (1984) Attachment of immunoglobulin to liposomal membrane via protein carbohydrate. Biochim Biophys Acta 800: 291
550. Heath TD, Macher BA, Papahadjopoulos D (1981) Covalent attachment of immunoglobulins to liposomes via glycosphingolipids. Biochim Biophys Acta 640: 66
551. Heath TD, Fraley RT, Papahadjopoulos D (1980) Antibody targeting of liposomes: cell specificity obtained by conjugation of F(ab')2 to vesicle surface. Science 210: 539
552. Barbet J, Machy P, Leserman LD (1981) Monoclonal antibody covalently coupled to liposomes: specific targeting to cells. J Supramol Struct Cell Biochem 16: 243
553. Martin FJ, Hubbell WL, Papahadjopoulos D (1981) Immunospecific targeting of liposomes to cells: a novel and efficient method for covalent attachment of Fab' fragments via disulfide bonds. Biochemistry 20: 4229
554. Martin FJ, Papahadjopoulos D (1982) Irreversible coupling of immunoglobulin fragments to preformed vesicles. J Biol Chem 257: 286
555. Hashimoto Y, Sugawara M, Endoh H (1983) Coating of liposomes with subunits of monoclonal IgM antibody and targeting of the liposomes. J Immunol Methods 62: 155
556. Derksen JTP, Scherphof GL (1985) An improved method for the covalent coupling of proteins to liposomes. Biochim Biophys Acta 814: 151
557. Snyder SL, Vannier WE (1984) Immunologic response to protein immobilized on the surface of liposomes via covalent azo-bonding. Biochim Biophys Acta 772: 288
558. Singer SJ (1974) Molecular organization of membranes. Annu Rev Biochem 43: 805
559. Tanford C (1978) Hydrophobic effect and organization of living matter. Science 200: 1012
560. Sinha D, Karush F (1979) Attachment to membranes of exogenous immunoglobulin conjugated to a hydrophobic anchor. Biochem Biophys Res Commun 90: 554
561. Sinha D, Karush F (1982) Specific reactivity of lipid vesicles conjugated with oriented anti-lactose antibody fragments. Biochim Biophys Acta 684: 187
562. Torchilin VP, Klibanov AL (1980) Preliminary "hydrophobization" of hydrophylic protein increases its binding with liposomes. Bioorg Khim (Russ.) 6: 791
563. Torchilin VP, Omel'yanenko VG, Klibanov AL, Mikhailov AI, Goldanskii VI, Smirnov VN (1981) Incorporation of hydrophylic protein modified with hydrophobic agent into liposome membrane. Biochim Biophys Acta 602: 511
564. Levashov AV, Kabanov AV, Khmelnitskii Yu L, Berezin IV, Martinek K (1984) Chemical modification of proteins (enzymes) by water-insoluble reagents. DAN SSSR (Russ) 278: 246
565. Torchilin VP, Klibanov AL, Smirnov VN (1982) Phosphatidylinositol may serve as the hydrophobic anchor for immobilization of protein on liposome surface. FEBS Lett 138: 117
566. Jansons VK, Mallett PL (1981) Targeted liposome: a method for preparation and analysis. Anal Biochem 111: 54
567. Huang A, Huang L, Kennell SJ (1980) Monoclonal antibody covalently coupled with fatty acid. J Biol Chem 225: 8015
568. Huang A, Tsao YS, Kennell SJ, Huang L (1982) Characterization of antibody covalently bound to liposomes. Biochim Biophys Acta 716: 140
569. Huang A, Kennell SJ, Huang L (1981) Immunoliposome labeling, a sensitive and specific method for cell surface labeling. J Immunol Methods 46: 141
570. Shen D-F, Huang A, Huang L (1982) An improved method for covalent attachment of antibody to liposomes. Biochim Biophys Acta 689: 31
571. Weissig V, Lasch J, Klibanov AL, Torchilin VP (1986) A new hydrophobic anchor for the attachment of proteins to liposomal membranes. FEBS Lett 202: 86

572. Torchilin VP, Ivanov NN, Klibanov AL, Smirnov VN (1979) Comparative study of different way of protein binding with liposomes. In 4th All union Biochemical congr Vol. 2, abstracts of scientific reports. Nauka, Moscow, p 131
573. Bogdanov AA, Klibanov AL, Torchilin VP (1984) Immobilization of α-chymotryspin on sucrose stearatepalmitate containing liposomes. FEBS Lett 175(1): 178
574. Wolf DE, Schlesseinger J, Elson IL, Webb WW, Blumenthal R, Henkart P (1977) Diffusion and patching of macromolecules on planar lipid bilayer membranes. Biochemistry 16: 3476
575. Klibanov AL (1984) Protein binding with liposomes. Ph. D. thesis. USSR cardiology research centre, Moscow.
576. Klibanov AL, Ivanov NN, Muzykantov VR, Torchilin VP (1985) The evaluation of quantitative parameters of antibody-bearing liposomes interaction with target antigens. Anal Bioch 150: 251
577. Rennard SI, Berg R, Martin GR, Foidart JM, Robey PG (1980) Enzyme-linked immunoassay (ELISA) for connective tissue components. Anal Biochem 104: 205
578. Machy P, Lesserman LD (1983) Small liposomes are better than large liposomes for specific drug delivery in vitro. Biochim Biophys Acta 720: 313
579. Gregoridis G, Meehan A (1981) Interaction of antibody-bearing small unilamellar liposomes with antigen-coated cells. Biochem J 200: 211
580. Margolis LB, Dorfman NA (1977) Preparation of liposomes possessing immunological specificity. Bull Exper Biol Med (Russ) 83: 53
581. Weinstein JM, Blumenthal R, Sharrow SO, Henkart PA (1978) Antibody mediated targeting of liposomes. Binding to lymphocytes does not ensure incorporation of vesicle contents into the cells. Biochim Biophys Acta 509: 272
582. Leserman LD, Machy P, Barbet J (1981) Cell specific drug transfer from liposomes bearing monoclonal antibodies. Nature (London) 293: 226
583. Harsh M, Walther P, Weder HG (1981) Targeting of monoclonal antibody coated liposomes to sheep red blood cells. Biochem Biophys Res Commun 103: 1069
584. Bragman KS, Heath TD, Papahadjopoulos D (1983) Simultaneous inteaction of monoclonal antibody-targeted liposomes with two receptors on K 562 cells. Biochim Biophys Acta 730: 187
585. Heath TD, Montgomery JA, Piper JR, Papahadjopoulos D (1983) Antibody-targeted liposomes: increase in specific toxicity of methotrexate-γ-aspartate. Proc Natl Acad Sci USA 80: 1377
586. Machy P, Barbet J, Leserman LD (1982) Differential endocytosis of T and B lymphocyte surface molecules evaluated with antibody-bearing fluorescent liposomes containing methotrexate. Proc Natl Acad Sci USA 79: 4148
587. Hashimoto Y, Sugawara M, Masuko T, Hojo H (1983) Antitumor effect of actinomycin D entrapped in liposomes bearing subunits of tumor-specific monoclonal IgM antibody. Cancer Res 43: 5329
588. Wolf B, Gregoriadis G (1984) The use of monoclonal anti-$Thy_1 IgG_1$ for the targeting of liposomes to AKP-A cells in vitro and in vivo. Biochim Biophys Acta 802: 259
589. Singhal A, Bali A, Gupta CM (1986) Antibody-mediated targeting of liposomes to erythrocytes in whole blood. Biochim Biophys Acta 880: 72
590. Norley SG, Huang L, Rouse BT (1986) Targeting of drug loaded immunoliposomes to herpes simplex virus infected corneal cells: an effective means of inhibiting virus replication in vitro. J Immunol 136: 681
591. Janah MM, Cerny EA, Rahman YE (1978) Tissue distribution of EDTA encapsulated within liposomes containing glycolipids or brain phospholipids. Biochim Biophys Acta 541: 321
592. Mauk MR, Gamble RC, Baldeschwieler JD (1980) Vesicle targeting timed release and specificity for leukocytes in mice by subcutaneous injection. Science 207: 309
593. Mauk MR, Gamble RC, Baldeschwieler JD (1980) Targeting of lipid vesicles specificity of carbohydrate receptor analogues for leukocytes in mice. Proc Natl Acad Sci USA, p 1430
594. Gregoriadis G, Neerunjun ED (1975) Homing of liposomes to target cells. Biochem Biophys Res Commun 65: 537
595. Ghosh P, Das PK, Bachhawat BK (1982) Targeting of liposomes towards different cell types of rat liver through the involvement of liposomal surface glycosides. Arch Biochem Biophys 213: 266
596. Rahman YE, Cerny EA, Patel KR, Lau EH, Wright BJ (1982) Differential uptake of liposomes varying in size and lipid composition by parenchymal and Kupfer cells of mouse liver. Life Sci 31: 2061

597. Das PK, Murray GJ, Zirzow GC, Brady RO, Barranger JA (1985) Lectin-specific targeting of β-glucocerebrosidase to different liver cells via glycosylated liposomes. Biochem Med 33: 124
598. Spanjer HH, Scherphof GL (1983) Targeting of lactosylceramide containing liposomes to hepatocytes in vivo. Biochim Biophys Acta 734: 40
599. Soriano P, Dijkstra J, Legrand A, Spanjer H, Londos-Gagliardi D, Roerdink F, Scherphof G, Nicolau C (1983) Targeted and nontargeted liposomes for in vivo transfer to rat liver cells of a plasmid containing the preproinsulin I gene. Proc Natl Acad Sci USA 80: 7128
600. Sunamoto J, Ywamoto K, Takada M, Yuzuriha T, Katayama K (1984) Improved drug delivery to target specific organs using liposomes as coated with polysaccharides. In: Chiellini E, Giusti P (eds) Polymers in medicine. Plenum, New York, p 157
601. Takada M, Yuzuriha Y, Katayama K, Iwamoto K, Sunamoto J (1984) Increased lung uptake of liposomes coated with polysaccharides. 802: 237
602. Poste G (1983) Liposome targeting in vivo: problems and opportunities. Biol Cell 47: 19
603. Dunnick JK, Badger RS, Takeda Y, Kriss JP (1976) Vasicles interactions with antibody and peptide hormone: role of vesicle composition. J Nucl Med 17: 1073
604. Weissmann G, Brand A, Franklin EC (1974) Interaction of immunoglobulins with liposomes. J Clin Invest 53: 536
605. Kaplan JH (1972) Anion diffusion across artificial lipid membranes: the effects of lysozyme on anion diffusion from phospholipid liposomes. Biochim Biophys Acta 290: 339
606. Juliano RL, Stamp D (1976) Lectin-mediated attachment of glycoprotein-bearing liposomes to cells. Nature (London) 261: 235
607. Begent RHJ, Keep PA, Green AJ, Searle F, Bagshawe KD, Jewkes RF, Jones BE, Barratt GM, Ryman BE (1982) Liposomally entrapped second antibody improves tumour imaging with radiolabelled (first) antitumour antibody. Lancet 2: 739
608. Campbell PJ, Harding NGL, Ryman BE, Tyreell DA (1980) Redistribution and altered excretion of digoxin in rats receiving digoxin antibodies incorporated in liposomes. Eur J Biochem 109: 87
609. Robinson PJ, Dunnill P, Lilly MD (1973) The properties of magnetic supports in relation to immobilized enzyme reactors. Biotechnol Bioeng 15: 603
610. Dunnill P, Lilly MD (1974) Purification of enzymes using magnetic bio-affinity materials. Biotechnol Bioeng 16: 987
611. Halling PI, Dunnill P (1980) Magnetic supports for immobilized enzymes and bioaffinity adsorbents. Enzyme Microb Technol 2: 2
612. Molday RS, Mackenzie D (1982) Immunospecific ferromagnetic iron-dextran reagents for the labeling and magnetic separation of cells. J Immunol Methods 52: 353
613. Kato T (1983) Encapsulated drugs in targeted cancer therapy. In: Bruck SD (ed) Controlled drug delivery, Vol. 2. CRC Press, Boca Raton, p 189
614. Kato T, Nemoto R, Mori H, Unno K, Goto A, Homma M (1980) An approach to magnetically controlled cancer chemotherapy. I. Preparation and properties of ferromagnetic mitomycin C microcapsules. J Jpn Soc Cancer Ther 15(5): 58
615. Kato T, Nemoto R, Mori H, Abe R, Unno K, Goto A, Murota H, Harada M, Kawamura K, Homma M (1981) An approach to magnetically controlled cancer chemotherapy. V. Preparation, properties and pharmacokinetics of microcapsules encased with magnetic particles and mitomycin C. J Jpn Soc Cancer Ther 16(7): 1351
616. Widder KJ, Marino PA, Morris RM, Senyei AE (1983) Targeting antineoplastic agents using magnetic albumin microspheres. In: Goldberg EP (ed) Targeting drugs. John Wiley and Sons, New York, p 201
617. Senyei AE, Driscoll CF, Widder KJ (1985) Biophysical drug targeting: magnetically responsive albumin microspheres. In: Widder KJ, Green R (eds) Drug and enzyme targeting, part a. Meth Enzymol 112, Academic Press, Orlando, p 56
618. Mosbach K, Schroder U (1979) Preparation and application of magnetic polymers for targeting of drugs. FEBS Lett 102: 112
619. Papisov MI, Torchilin VP (1987) Magnetic drug targeting. II. Targeted drug transport by magnetic microparticles: factors influencing therapeutic effect. Int J Pharm 40: 207
620. Papisov MI, Samokhin GP, Smirnov MD, Torchilin VP, Smirnov VN (1984) Possibilities of using ferromagnetic materials for targeted drug transport. Bull Exper Biol Med (Russ) 98(9): 372
621. Widder KJ, Senyei AE, Scarpelli DG (1978) Magnetic microspheres: a model system for site specific drug delivery in vivo. Proc Soc Exper Biol Med 58: 141

622. Chaplin MF, Kennedy JF (1976) Magnetic, immobilized derivatives of enzymes. Carbohydr Res 50: 267
623. Torchilin VP, Papisov MI, Smirnov VN (1985) Magnetic Sephadex as a carrier for enzyme immobilization and drug targeting. J Biomed Mater Res 19: 461
624. Kandzia J, Scholz W, Anderson MJD, Muller-Ruchholtz W (1984) Magnetic albumin/protein A immunomicrospheres. I. Preparation, antibody binding capacity and chemical stability. J Immunol Meth 75: 31
625. Burns MA, Kvesitadze GI, Graves DJ (1985) Dried calcium alginate/magnetite spheres: a new support for chromatographic separations and enzyme immobilization. Biotechnol Bioeng 27: 137
626. Kiwada H, Sato J, Yamada S, Kato Y (1986) Feasibility of magnetic liposomes as a targeting device for drugs. Chem Pharm Bull 34(10): 4253
627. Rusetsky AN, Papisov MI, Ruuge EK, Torchilin VP (1985) On the pre-conditions of using magnet-directed localization of drugs for thrombosis treatment. Bull VKNTs (Bull USSR Card Res Ctr Russ) 1: 100
628. Ruuge EK, Rusetsky AN (1987) Targeted drug transport by magnetic field. USSR Mendeleev Chem J (Russ.) 5: 556
629. Torchilin VP, Papisov MI, Orekhova NM, Belyaev AA, Petrov AD, Ragimov SE (1988) Magnetically driven thrombolytic preparation containing immobilized streptokinase. Targeted transport and action. Hemostasis 18: 113
630. Senatore F, Bernath F, Meisner K (1986) Clinical study of urokinase-bound fibrocollagenous tubes. J Biomed Mater Res 20: 177
631. Senatore F, Bernath F (1986) Lysis time and FDP immunoprecepitation by soluble and immobilized urokinase. J Biomed Mater Res 20: 189
632. Senotore FF, Bernath FR (1986) Urokinase bound to fibrocollagenous tubes: an in vitro kinetic study. Biotechnol Bioeng 28: 58
633. Senatore F, Bernath F (1986) Parameters affecting optimum activity and stability of urokinase-bound fibrocollagenous tubes. Biotechnol Bioeng 28: 64
634. Solomon DD, McGary CW, Pascarella VJ (1985) Covalent bounded antithrombogenic polyurethane material. US Patent 4 521 564
635. Kim SW (1983) The biological activity of antithrombic agents, immobilized on polymer surfaces. Ann NY Acad Sci 416: 513
636. Kim HP, Byun SM, Yeom YI, Kim SW (1983) Immobilization of urokinase on agarose matrices. J Pharm Sci 72(3): 225
637. Oshiro T, Supitachi A, Hong SD, Mukai K, Murakami K, Kosaki G, Motoi I (1978) Immobilization of fibrinolytic enzymes. Fibrinolytic activity of immobilized urokinase linked to Sepharose 6B. Blood and Vessel (Jap.) 9: 72
638. Sugitachi A, Kawahara T, Kodama J, Kikkawa Y, Takagi K (1978) Antithrombogenicity of immobilized urokinase and its clinical significance. Blut 37: 31
639. Oshiro T, Kosaki G (1980) Urokinase immobilized on medical polymeric materials: fundamental and clinical studies. Artific Organs (Jap.) 4: 58
640. Oshiro T, Takahashi A, Mukai K, Hong DS, Kosaki G, Tagaki K (1978) The study on available immobilization method of urokinase to nylon. Artific Organs (Jap.) 2: 210
641. Klein MD, Langer R (July 1986) Immobilized enzymes in clinical medicine: an emerging approach to new drug therapies. TIBTECH p 179
642. Pedersen H, Horvath C (1982) Pharmacokinetic characterization of extracorporeal therapy. J Pharmacokinet Biopharm 10: 437
643. Bertino JR, Condos S, Horvath C, Kalghatgi K, Pedersen H (1978) Immobilized carboxypeptidase G in methotrexate removal. Cancer Res 38: 1936
644. Bulygin AN Klyosov AA, Rostchin SA, Bulygina LA, Pasechnik VA, Khalyapin BD, Shtukina TV (1984) Mathematical modelling of tubular extracorporeal shunts with immobilized enzymes. Prikl Biochim Microbiol (Russ) 20: 230
645. Salmona M, Saronio C, Garattini S (eds) (1974) Insolubilized enzymes. Raven Press, New York
646. Ko RYC, Hersh LS (1976) A dacron wool packed-bed extracorporeal reactor: a kinetic study of immobilized Escherichia coli II L-asparaginase. J Biomed Mater Res 10: 249
647. Horvath C, Sardi A, Woods JS (1973) L-asparaginase tubes: kinetic behavior and application in physiological studies. J Appl Physiol 34: 181
648. Olanoff LS, Bernath FR, Venkatasubramanian K (1975) Perfusion trials with collagen immobililzed enzyme in an extracorporeal reactor. Amer Chem Soc Polym Prepr 16: 203

649. Sampson D, Han T, Hersh LS, Murphy GP (1974) Extracorporeal chemotherapy with L-asparaginase in man. J Surg Oncol 6: 39
650. Mazzola G, Vecchio G (1980) Immobilization and characterization of L-asparaginase on hollow fibers. Int J Artif Organs 3: 120
651. Bunting PS, Laidler KJ (1974) Flow kinetics of L-asparaginase attached to nylon tubing. Biotechnol Bioeng 16: 119
652. Khalyapin BD, Shtukina TV, Voinova NE, Pasechnik VA (1980) The development of extracorporeal shunt with immobilized L-asparaginase. In: abstracts of III USSR conference on enzyme engineering. Leningrad, p 208
653. Gombotz WR, Hoffman AS, Schmer G, Uenoyama S (1985) The immobilization of L-asparaginase on porous hollow fiber plasma filters. J Contr Rel 2: 375
654. Dumler F, Singh P, Jackson CE, Kini KR, Samhouri AM, Halvorson HR, Shore JD (1981) Extracorporeal enzyme therapy. Use of an L-asparaginase reactor-dialyzer in a clinical setting. Am Soc Artif Int Org J 4(2): 70
655. Rossi V (1981) Immobilization of arginase on hollow-fiber hemodializer. Int J Artif Organs 4(2): 102
656. Kanalas JJ, Spector EB, Cederbaum SD (1982) Hollow-fiber reactors containing mammalian arginase: an approach to enzyme replacement therapy. Biochem Med 27: 46
657. Kalghatgi KK, Horvath C, Ambrus CM (1980) Multi-tubular reactors with immobilized L-phenylalanine ammonia lyase for use in extracorporeal shunts. Res Commun Chem Pathol Pharmacol 27: 551
658. Pedersen H, Horvath C, Ambrus CM (1979) Preparation of immobilized L-phenylalanine ammonia lyase in tubular forms for depletion of L-phenylalanine. Res Commun Chem Pathol Pharmacol 20: 559
659. Ambrus JJ, Codey C, Virand EA (1983) In vivo safety of hollow-fiber enzyme-reactors with immobilized phenylalanine ammonia lyase in a large animal model for phenylketonurea. J Pharmacol Exp Ther 224: 598
660. Ambrus CM (1982) Depletion of phenylalanine in the blood of phenylketonuric patients using a PAL-enzyme reactor. An in vitro study. Res Commun Chem Pathol Pharmacol 37: 105
661. Larue C, Munnick A, Charpentier C, Sandubray JM, Frezal J, Romy MJ, Rivat C (1986) An extracorporeal hollow-fiber reactor for phenylalanineketoneurea using immobilized phenylalanine ammonia lyase. Develop pharmacol Ther 9(2): 73
662. Ambrus CM, Ambrus JJ, Kant C, Mirand EA, Sharma SD, Horvath CS, Pedersen H, Guthrie R, Paul TD (1978) Phenylalanine depletion for the management of phenylketonurea. Use of enzyme reactors with immobilized enzymes. Science 201: 837
663. Salmona M, Saronio C, Bartosek I, Vecchi A, Mussini E (1974) Potential biomedical applications of fiber-entrapped L-asparaginase and urease. In: Salmona M, Saronio C, Garattini S, (eds) Insolubilized enzymes. Raven Press, New York, p 189
664. Makarov KA, Kibardin SA (1980) Immobilized biopreparations in medicine. Meditsina (Russ) Moscow
665. Mel'nik IP, Molodenkov MN, Mashkov OA, et al. (1977) Study of immobilized urease use for the decomposition of urea in blood plasma. Bull Exper Biol Med (Russ) 83: 425
666. Ota K (1974) A new method of urea removal using urease and expanded polytetrafluoroethylene membrane. Artific Organs (Jap) 3: 350
667. Cherkasova TA, Yanina MM, Bogdanov VI, Kurganov BI (1983) Convertor with immobilized urease for urea decomposition in "artificial kidney" apparatus. In abstracts of IV USSR conference on enzyme engineering. Kiev, p 73
668. Bollmeier JP (1979) Hydrolysis of urea by gelatin-immobilized urease: separation of diffusion and kinetic phenomena in a model immobilized-enzyme reactor system. Biotechnol Bioeng 21: 2303
669. Davidenko TI, Sevast'yanova EV, Morozova AA, Andronati SA, Ermolenko IN, Yuodvalkite DYu, Glemza AA (1985) Urease adsorption on coal fibers. Doklady AN Ukr SSR 1: 61
670. Venter JC, Venter BR, Dixon JE, Kaplan NO (1975) A possible role for glass bead immobilized enzymes as therapeutic agents (immobilized uricase as enzyme therapy for hyperuricemia). Biochem Med 12: 79
671. Plotz PH, Berk PD, Scharschmidt BF, Gordon JK, Vergalla J (1974) Removing substances from blood by affinity chromatography. I. Removing bilirubin and other albumin-bound substances from plasma and blood with albumin-conjugated agarose beads. J Clin Invest 53: 778

672. Sung C, Lavin A, Klibanov AM, Langer R (1986) An immobilized enzyme reactor for the detoxification of bilirubin. Biotechnol Bioeng 28: 1531
673. Schmer G (1984) Biochemically active matrix for use in a bioartificial organ. US Patent 4 438 198
674. Langer R, Linhardt RJ, Hoftberg S, Larsen AK, Cooney CL, Tapper D, Klein M (1982) An enzymatic drug removal system for heparin in extracorporeal therapy. Science 217: 261
675. Linhardt RJ, Cooney CL, Tapper D, Zannetos CA, Larsen AK, Langer R (1984) An immobilized microbial heparinase for blood deheparinization. Appl Biochem Biotechnol 9: 41
676. Larsen AK, Linhardt RJ, Tapper D, Klein M, Langer R (1984) Effect of extracorporeal blood deheparinization on formed blood components. Artif Organs 8: 198
677. Hagiwara H, Nagasaki T, Saito Y, Inada Y (1982) Fibrin-membrane endowed with biological functions. 7. An approach to an artificial liver: conversion of ammonia to urea in vitro. Biochem Biophys Res Commun 104: 507
678. Okamoto H, Kanai S, Tipayang P, Inada Y (1979) Fibrin membrane endowed with biological function. IV. Formation of cross-links between fibrinogen (or fibrin) and ribonuclease by transglutaminase. Enzyme 24: 273
679. Siegbahn N, Mosbach K (1982) An immobilized cyclic multistep enzyme system—the urea cycle. FEBS Lett. 137: 6
680. Inada Y, Tazawa Y, Attygalle A, Saito Y (1980) Fibrin membrane endowed with biological function VI. Immobilization of multienzymes of urea cycle. Biochem Biophys Res Commun 96: 1586
681. Miura Y (1981) Detoxification of ammonia by immobilized urea cycle enzymes. Artif Organs 5: 72
682. Plotz PH, Berk PD, Scharschmidt BF, Gordon JK, Vergallal J (1974) Removing substances from blood by affinity chromatography. I. Removing bilirubin and other albumin-bound substances from plasma and blood with albumin-conjugated agarose beads. J Clin Invest 53: 778
683. Scharschmidt BF, Plotz PH, Berk PD, Waggoner JG, Vergalla J (1974) Removing substances from blood by affinity chromatography. II. Removing bilirubin from the blood of jaundiced rats by hemoperfusion over albumin-conjugated agarose beads. J Clin Invest 53: 786
684. Valuev LI (1988) Macromonomers of physiologically active compounds. VINITI, Moscow
685. Valueva TA, Maklakova IA, Valuev LI, Mosolov VV, Plate NA (1985) Biospecific sorbent to remove proteases from biological fluids. Vopr Med Khim (Russ 31(4): 34
686. Larionova NI, Mityushina GV, Vladimirov VG, Zhuravlev AG (1983) Biospecific matrices containing of natural and synthetic inhibitors of proteases. In: abstracts of IV USSR conference on enzyme engineering. Kiev, vol. 1: 110
687. Siu Chong ED, Chang TMS (1977) L-Asparaginase as a model for enzyme therapy of substrate-dependent tumors. In: Chang TMS (ed) Biomedical applications of immobilized enzymes and proteins, v. 1. Plenum Press, New York, p 105
688. Chang TMS, Migchelsen M (1973) Characterization of possible "toxic" metabolites in uremia and hepatic coma based on the clearance spectrum for larger molecules by the ACAC microcapsule artificial kidney. Trans Amer Soc Artif Int Organs 19: 314
689. Chang TMS, Malouf C (1980) Effects of glucose dehydrogenase in converting urea and ammonia into aminoacid using artificial cells. Artif Organs 3: 38
690. Chang TMS (1977) Artificial kidney, artificial liver, and detoxifiers based on artificial cells, immobilized proteins, and immobilized ensymes. In: Chang TMS (ed) Biomedical applications of immobilized enzymes and proteins. New York, p 281
691. Ortmains A, Neufeld RY, Chang TMS (1984) Study of microencapsulated urease in a continuous feed, stirred, tank reactor. Enz Microb Technol 6: 135
692. Chang TMS (1980) New approaches using immobilized enzymes for the removal of urea and ammonia. Weetall H (ed) Enzyme Eng, v. 5. Plenum Press, New York, p 225
693. Terman DS, Tavel A, Tavel T, Petty D, Harbeck R, Buffaloe G, Carr R (1976) Degradation of circulating DNA by extracorporeal circulation over nuclease immobilized on nylon microcapsules. J Clin Invest 57: 1201
694. Virnik AD (1985) Chemistry of medical fibers and textiles: achievements and problems. J V Kh O (Russ.) 30: 447
695. Volf LA (1980) Fibers with special properties. Khimia (Chemistry Press, Russ) Moscow
696. Proceedings of International Symposium on Chemical Fibers. Kalinin, 1981, v. 5

697. Kil'deeva NR, Larionova NI, Kazanskaya NF, Virnik AD (1980) Immobilization of modified α-chymotrypsin within the structure of cellulose triacetate membranes. Biokhimia (Russ) 45: 569
698. Virnik AD, Ryltsev VV, Vlasov LG, Dovbii EV, Kalashnik AT (1983) Effect of γ-radiation on trypsin immobilized on dialdehyde cellulose. Prikl Biokhim Microbiol (Russ) 19: 533
699. Linko YY, Poljola L, Viskari R, Linko M (1976) A simple entrapment method for immobilizing enzymes within cellulose fibers. FEBS Lett 62: 77
700. Chuenko AV, Davidenko TI, Potapskii YuP, Bondarchuk AA, Andronati SA, Zelezny VI, Zukovskii YuM (1986) Immobilization of proteolytic enzymes on dressing materials. Doklady AN Ukr SSR (Russ) Series B 1: 81
701. Gostishchev VK, Vasilkova ZF, Vavilova GS, Vladimirov VG, Virnik AD, Larionova NI (1982) The application of immobilized proteolytic enzymes and their inhibitors in surgery. Vestnik Khirurgii (Russ) 129: 89
702. Tolstykh PI, Arutyunyan BN, Struchkov YuV, Vasilkova ZF, Krasovskaya SB, Belyaeva DA, Kildeeva NR (1980) Biologically active suture material as a measure for prevention of disturbed wounds healing (Literature survey) Khirurgia (Russ) 5: 108
703. Tolstykh PI, Gostistchev VK, Vasil'kova ZF, Vlasov LG, Rul'tsev VV, Samoilova TI, Sadykov AG, Yudanova TN, Ilyuk VV, Karpov GS (1985) Textile immobilized proteases in therapy of pus-necrotic deseases of soft tissues. Abstracts of V USSR conference on enzyme engineering, Kobuleti, p 128
704. Stepanova LS, Kudryavtseva TV, Shamolina II, Volf LA (1980) Use of enzyme-containing fibers in therapeutic stomatology. Abstracts of III USSR conference on enzyme engineering, Leningrad, p 230
705. Gonchar AM, Kogan AC, Salganik PI (1986) Wound healing and immobilized proteolytic enzymes. Nauka (Russ) Moscow
706. Petchot-Bacque JP, Papantoniou C, Mahieu C, Shroot B (1984) Materiau polymerique a action enzymatique, son procede de preparation et son utilisation en tant que medicament. French Patent 2 558 171
707. Guilbault GG, de Olivera Neto G (1985) Immobilized enzyme electrodes. In: Woodward J (ed) Immobilized cells and enzymes. A practical approach. IRL Press, Oxford, p 55
708. Guilbalt G (1984) Handbook of immobilized enzymes. Marcel Dekkev, New York
709. Busby MG, Hartwig DE (1982) Glucose analyser membrane containing immobilized glucose oxidase. US Patent 4 317 879
710. Lange MA, Chambers JQ (1985) Amperometric determination of glucose with a ferrocene-mediated glucose oxidase/polyacrylamide gel electrode. Anal Chim Acta 175: 89
711. Mell LD, Maloy JT (1975) A model for the amperometric enzyme electrode obtained through digital simulation and applied to the immobilized glucose oxidase system. Anal Chem 47: 299
712. Wingard Jr LB, Ellis D, Yao SJ, Schiller JG, Liu CC, Wolfson JrSK, Drash AL (1979) Direct coupling of glucose oxidase to platinum and possible use for in vivo glucose determination. J Solid Phase Biochem 4: 253
713. Wingard LB Jr Cantin LA, Castner JF (1983) Effect of enzyme-matrix composition on potentiometric response to glucose using glucose oxidase immobilized on platinum. Biochim Biophys Acta 748: 21
714. Wingard LB (1983) Immobilized enzyme electrodes for glucose determination for the artificial pancreas. Fed Proc 42: 2
715. Wingard Jr LB, Castner JF (1982) Potentiometric glucose sensor: enzymatic activity and potentiometric measurements. In: Chibata Y, Fukui S, Wingard LB Jr (eds) Enzyme engineering. Plenum Press, New York, p 415
716. Wingard Jr LB, Castner JF, Miyawaki O, Marrese C (1984) Rotating ring-disc electrode: its use in studying mass transport resistances through immobilized enzyme support matrices. Ann NY Acad Sci 434: 224
717. Castner JF, Wingard LB Jr (1984) Alterations in potentiometric response of glucose oxidase platinum electrodes resulting from electrochemical or thermal pretreatments of a metal surface. Anal Chem 56: 2891
718. Wingard LB Jr, Castner JF, Yao SJ, Wolfson SK Jr, Drash AL, Liu CC (1984) Immobilized glucose oxidase in the potentiometric detection of glucose. Appl Biochem Biotechnol 9: 95
719. Bertermann K, Elze P, Scheller F, Pfeiffer D, Janehen M (1982) The application of an enzyme electrode for blood glucose determination in the automated flow system. Anal Lett 16(B 15): 397

720. Castner JF, Wingard Jr LB (1981) Mass transport and reaction kinetic parameters determined electrochemically for immobilized glucose oxidase. Biochemistry 23: 2203
721. Mullen WH, Churchhouse SJ, Keedy FH, Vadgama PM (1986) Blood glucose determination using an enzyme electrode based on the quinoprotein, glucose dehydrogenase. Anal Proc 23: 145
722. Ugarova NN (1982) The use of enzymes in chemical analysis. In: Berezin IV, Martinek K (eds) Introductory course of applied enzymology. MGU, Moscow, p 306
723. Llenado RA, Rechnitz GA (1974) Automated serum urea determination using membrane electrodes. Anal Chem 46: 1109
724. Mascini M, Guibault GG (1977) Urease coupled ammonia electrode for urea determination in blood serum. Anal Chem 49: 795
725. Adams RE, Carr PW (1978) Coulometric flow analyzer for use with immobilized enzyme reactors. Anal Chem 50: 944
726. Nanjo M, Guilbault GG (1974) Enzyme electrode sensing oxygen for uric acid in serum and urine. Anal Chem 46: 1769
727. Papastathopoulos DS, Rechnitz GA (1975) Enzymatic cholesterol determination using ion-selective membrane electrodes. Anal Chem 47: 1792
728. Cullen LF, Rusling JF, Schleifer A, Papariello GJ (1974) Improved penicillin selective enzyme electrode. Anal Chem 46: 1955
729. Posadka P, Macholan L (1979) Amperometric assay of vitamin C using an ascorbate oxidase enzyme electrode. Collect Czech Chem Commun 44: 3395
730. Nady J, Rice ME, Adams PN (1982) A new type of enzyme electrode: the ascorbic acid eliminator electrode. Life Sci 31: 2611
731. Vincke BJ, Devleeschouwer MJ, Patriarche GJ (1985) Determination de l'acide L-ascorbique a l'acid d'electrodes bacteriennes, tissulaires et enzymatiques. Anal Lett 18(B 13): 1593
732. Ciucu A, Magearu V, Luca C (1985) Enzyme electrode sensor for hydrogen peroxide. Anal Lett 18(B 3): 299
733. Bradley CR, Rechnitz GA (1988) Comparison of oxalate oxidase enzyme electrodes for urinary oxalate determinations. Anal Lett 19: 151
734. Kerube J, Matsunaga T, Teraoka N, Suzuki S (1980) Microbioassay of phenylalanine in blood sera with a lactate electrode. Anal Chem Acta 119: 271
735. Tsuchida T, Yoda K (1981) Enzyme electrodes for determination of creatinine and creatine in serum. J Clin Chem Clin Biochem 19: 861
736. Tsuchida T, Yoda K (1983) Multi-enzyme membrane electrodes for determination of creatinine and creatine in serum. Clin Chem 29: 51
737. Wingard Jr LB (1984) Cofactor modified electrodes. Trends Anal Chem 3: 235
738. Narasimhan K, Wingard Jr LB (1986) Immobilization of flavins on electrode surfaces. I. Attachment of riboflavin through 8-methyl to glassy carbon. J Molec Catal 34: 253
739. Narasimhan K, Wingard Jr LB (1986) Immobilization of flavins on electrode surfaces. II. Apparent conversion of glassy carbon-immobilized riboflavin to immobilized flavin adenin dinucleotide and partial reconstitution of apoglucose oxidase activity. J Molec Catal 34: 263
740. Niiyama Y, Sugahara K (1986) Electrochemical sensor having and immobilized enzyme membrane. US Patent 4 579 642
741. Foulds NC, Lowe CR (1986) Enzyme entrapment in electrically conducting polymers. J Chem Soc Faraday Trans 82: 1259
742. Frew JE, Green MJ, Hill HAO (1986) New electrochemical techniques applied to medicine and biology. J Chem Tech Biotechnol 36: 357
743. Lomen CE, de Alwis U, Wilson GS (1986) Use of a reversibly immobilized enzyme in the flow injection—amperometric determination of picmole glucose levels. J Chem Soc Faraday Trans 82: 1265
744. Dzantiev BB, Egorov AM (1982) Current state and the development prospects of the enzyme immunoassay. Zhurn V Kh O (Russ.) 22: 82
745. Sittampalam GS, Wilson GS (1984) Enzyme immunoassays with electrochemical detection. Trends Anal Chem 3: 96
746. Guesdon JL, Avrameas S (1981) Solid phase immunoassays. Appl Biochem Bioeng 3: 207
747. Ngo TT, Lenhoff HM (1982) Enzymes as versatile labels and signal amplifiers for monitoring immunochemical reactions. Mol Cell Biochem 44: 3
748. Ngo TT, Lenhoff HM (1981) Recent advances in homogeneous and separation-free enzyme immunoassays. Appl Biochem Biotechol 6: 53

748a. Ngo TT, Lenhoff HM (eds) (1985) Enzyme-mediated immunoassay. Plenum Press, New York and London
749. Dzantiev BV, Gavrilova EM, Egorov AM (1982) Enzyme immunoassay. In: Berezin IV, Martinek K (eds) Introductory course of applied enzymology. MGU, Moscow, p 343
750. van Weemen BK, Schuurs AHWM (1974) Immunoassay using antibody-enzyme conjugates. FEBS Lett 43: 215
751. Belanger L, Hamel D, Dufour D, Pouliot M (1976) Double-antibody enzyme immunoassay applied to human alpha 1-fetoprotein. Clin Chem 22: 198
752. Maiolini R, Masseyeff R (1975) A sandwich method of enzymoimmunoassay. I. Application to rat and human alpha-fetoprotein. J Immunol Meth 8: 223
753. Ishikawa E (1973) Enzyme immunoassay of insulin by fluorimetry of the insulin-glucoamylase complex. J Biochem 73: 1319
754. Kennedy JH, Kricka LJ, Wilding P (1976) Protein-protein coupling reactions and the applications of protein conjugates. Clin Chim Acta 70: 1
755. Kato K, Hamaguchi Y, Fukui H, Ishikawa E (1975) Enzyme-linked immunoassay. II. A simple method for synthesis of the rabbit antibody-beta-D-galactosidase complex and its general applicability. J Biochem 78: 423
756. Kato K, Hamaguchi Y, Fukui H, Ishikawa E (1975) Enzyme-linked immunoassay. I. Novel method for synthesis of the insulin-beta-D-galactosidase conjugate and its applicability for insulin assay. J Biochem 78: 235
757. Kitagawa T, Aikawa T (1976) Enzyme coupled immunoassay of insulin using a novel coupling reagent. J Biochem 79: 233
758. Ternynck T, Avrameas S (1977) Conjugation of p-benzoquinone treated enzyme with antibodies and Fab fragments. Immunochemistry 14: 767
759. Pillai S, Bachhawat BK (1978) Monoconjugate enzyme linked immunoassay. FEBS Lett 90: 51
760. O'Sullivan MJ, Gnemmi E, Morris D, Chieregatti G, Simmons M, Simmonds AD, Bridges JW, Marks V (1978) A simple method for the preparation of enzyme-antibody conjugates. FEBS Lett 95: 311
761. Yoshitake S, Yamada Y, Ishikawa E, Masseyeff R (1979) Conjugation of glucose oxidase from Aspergillus niger and rabbit antibodies using N-hydroxysuccinimide ester of N-(4-carboxycyclohexylmethyl)-maleimide. Eur J Biochem 101: 395
762. Nilsson P, Bergquist NR, Grundy MS (1981) A technique for preparing defined conjugates of horseradish peroxidase and immunoglobulin. Immunol Meth 41: 81
763. Karaseva EI, Eremin AN, Metelytsa DI (1987) Preparing and characterization of glucose-6-phosphate dehydrogenase with cortisol and progesteron. Biotekhnologia (Russ) 3: 198
764. Aikawa R, Ohyama K, Mori E, Hironaka K, Miyake M, Morishita M (1985) Enzyme immunoassay for elcatonin, a synthetic analoque of eel calcitonin, in plasma. J Pharm Sci 74: 300
765. Avrameas S (1969) Indirect immunoenzyme techniques for the intracellular detection of antigens. Immunochemistry 6: 825
766. Romano EL, Romano M (1977) Staphylococcal protein a bound to colloidal gold: a useful reagent to label antigen-antibody sites in electron microscopy. Immunochemistry 14: 711
767. Halliday MI, Wisdom GB (1978) A copetitive enzyme immunoassay using labelled antibody. FEBS Lett 96: 298
768. Dray F, Andrieu J-M, Renaud R (1975) Enzyme immunoassay of progesterone at the picogram level using galactosidase as label. Biochim Biophys Acta 403: 131
769. Carrico RJ, Yeung K-K, Schroeder HR, Boguslaski RC, Buckler RT, Christner JE (1976) Specific protein-binding reactions monitored with ligand-ATPase conjugates and firefly luciferase. Anal Biochem 76: 95
770. Mattiasson B, Nilsson H (1977) An enzyme immunoelectrode. Assay of human serum albumin and insulin. FEBS Lett 78: 251
771. Mattiasson B, Borrebaeck C, Sanfridson B, Mosbach K (1977) Thermometric enzyme linked immunosorbent assay: TELISA. Biochim Biophys Acta 483: 221
772. Mattiasson B (1977) A general enzyme thermistor based on specific reversible immobilization using the antigen-antibody interaction. Assay of hydrogen peroxide, penicillin, sucrose, glucose, phenol and tyrosine. FEBS Lett 77: 107
773. Haga M, Itagaki H, Sugawara S, Okano T (1980) Liposome immunosensor for theophylline. Biochem Biophys Res Commun 95: 187
774. Porstmann B, Portsmann T, Gaede D, Nugel E, Egger E (1981) Temperature dependent rise in activity of horseradish peroxidase caused by non-ionic detergents and its use in enzyme-immunoassay. Clin Chim Acta 109: 175

775. Kricka LJ, Thorpe GHG (1983) Chemiluminescent and bioluminescent methods in analytical chemistry. Analyst 108: 1274
776. Ugarova NN, Brovko LYu, Lebedeva OV (1986) Immobilized bioluminescent systems in analytical biochemistry and biotechnology. In: Berezin IV (ed) The use of immobilized enzymes. VINITI, Moscow, p 88
777. Brovko LYu, Ugarova NN, Vasileva TE, Dombrovsky VA, Berezin IV (1978) Administration of immobilized firefly luciferase for quantitative estimation of ATP and enyzmes synthetizing and destroying ATP. Biokhimia (Russ) 43: 798
778. Ugarova NN, Brovko LYu, Kost NV (1982) Immobilization of luciferase from the firefly Luciola mingrelica-catalytic properties and stability of the immobilized enzyme. Enzyme Microb Technol 4: 224
779. Wienhausen GK, Kricka LJ, Hinkley JE, De Luca M (1982) Properties of bacterial luciferase/NADH-FMN oxidoreductase and firefly luciferase immobilized into Sepharose. Appl Biochem Biotechnol 7: 463
780. Ugarova NN, Brovko IYu, Beliaieva EI (1983) Immobilization of lucife from the firefly Luciola mingrelica: catalytic properties and thermostability of the enzyme immobilized on cellulose films. Enzyme Microb Technol 5: 60
781. Lebedeva OV, Ugarova NN, Deyko TV, Raynina EI, Makhlis TA (1985) A bioluminescent method for determining picomolar quantities of nicotinamide adenine dinucleotide using an immobilized extract of the luminous bacteria Beneckea harveyi. Prikl Biokhim Microbiol (Russ) 21: 114
782. Rodriguez O, Guilbault GG (1981) Immobilized bacterial luciferase for microscale analysis of creatine kinase activity. Enzyme Microb Technol 3: 69
783. Kurkijarvi K, Raunio R, Korpela T (1982) Packed-bed reactor of immobilized bacterial bioluminescence enzyme: a potential high-sensitivity detector for automated analyzers. Anal Biochem 125: 415
784. Jablonski E, De Luca M (1979) Properties and uses of immobilized light-emitting enzyme systems from Beneckea harveyi. Clin Chem 25: 1622
785. Kricka LJ, Wienhausen GK, Hinkley JE, De Luca M (1983) Automated bioluminescent assays for NADH, glucose 6-phosphate, primary bile acids and ATP. Anal Biochem 129: 392
786. Mosbach K, Danielsson B, Borgerud A, Scott M (1975) Deterimation of heat changes in the proximity of immobilized enzymes with an enzyme thermistor and its use for the assay of metabolites. Biochim Biophys Acta 403: 256
787. James DR, Pring B (1975) Use of immobilized urease in a continuous flow method for the determination of plasma urea. Clin Chim Acta 62: 435
788. Sundaram PV, Igloi MP, Wassermann R, Hinsch W, Knoke KJ (1978) Immobilized-enzyme nylon-tube reactors for routine determination of urea and citrulline in serum. Clin Chem 24: 234
789. Karube I, Kubo I, Suzuki S (1984) Hybrid biosensor for clinical and fermentation process control. Ann NY Acad Sci 434: 508
790. Cunningham SK, Keaveny TV (1978) A two-stage enzymatic method for determination of uric acid and hypoxanthine/xanthine. Clin Chim Acta 86: 217
791. Kamoun P, Douay O (1978) Immobilized-enzyme membrane sandwich reactor used in automated micro-seale assay of plasma uric acid. Clin Chem 24: 2033
792. Salleh AB, Ledingham WM (1981) Uricase immobilized onto nylon tube for use in automated analysis of serum urate. Anal Biochem 116: 40
793. Iob A, Mottola HA (1981) Uricase immobilized on walls of glass tubes used as reactors in continuous-flow analyzes. Clin Chem 27: 195
794. Nakamura N, Murayama K, Kinoshita T (1986) Immobilization of uricase on protamine bound to glass beads and its application to determination of uric acid. Anal Biochem 152: 386
795. Sundaram PV (1982) Routine determination of hydrogen peroxide in clinical chemistry with immobilized aldehyde dehydrogenase. In: Chibata I, Fukui S, Wingard Jr LB (eds) Enzyme engineering 6. Plenum Press, New York, p 473
796. Postick DT, Hercules DM (1975) Quantitative determination of blood glucose using enzyme induced chemiluminescence of luminol. Anal Chem 47: 447
797. Leon LP, Narayanan S, Dellenhach R, Horvath C (1976) Immobilized glucose oxidase used in the continuous-flow determination of serum glucose. Clin Chem 22: 1017
798. Okuda J, Miwa I, Maeda K, Tokui K (1977) Rapid and sensitive, colorimetric determination of the anomers of D-glucose with D-glucose oxidase, peroxidase and mutarotase. Carbohydrate Res 58: 267

799. Tabata M, Fukunaga C, Ohyabu M, Murachi T (1984) Highly sensitive flow injection analysis of glucose and uric acid in serum using an immobilized enzyme column and chemiluminescence. J Appl Biochem 6: 251
800. Bovara R, Carrea G, Cremonesi P, Mazzola G (1981) Continuous-flow analysis of 3-alpha-hydroxysteroids using immobilized 3-alpha-hydroxysteroid dehydrogenase. Anal Biochem 112: 239
801. Day DF, Workman WE (1984) A simple insulin assay for renal clearance determination using an immobilized β-fructofuranosidase. Ann NY Acad Sci 434: 504
802. Murachi T (1982) Use of immobilized enzyme reactors in automated clinical analysis. In: Chibata I, Fukui S, Wingard Jr LB (eds) Enzyme engineering 6. Plenum Press, New York, p 369
803. Thompson RQ, Mandoke CS, Womack JP (1985) A Procedure for immobilizing enzymes on nylon. Anal Lett 18(B 1): 93
804. Radhavan KG, Ramakrishnan V (1986) Fabrication of reusable enzyme brushes for clinical analysis: lactate dehydrogenase enzyme brush. Biotechnol Bioeng 28: 1611
805. Radhavan KG, Devasagayam TPA, Ramakrishnan (1986) Immobilized-enzyme brushes for clinical analyses: urea determination. Anal Lett 19: 163

Author Index Volumes 1–11

The volume numbers are printed in Italics

MIX
Papier aus verantwortungsvollen Quellen
Paper from responsible sources
FSC® C105338
FSC
www.fsc.org

If you have any concerns about our products,
you can contact us on
ProductSafety@springernature.com

In case Publisher is established outside the EU,
the EU authorized representative is:
Springer Nature Customer Service Center GmbH
Europaplatz 3, 69115 Heidelberg, Germany

Printed by Libri Plureos GmbH
in Hamburg, Germany